ADVANCES IN
EDUCATIONAL ADMINISTRATION

Volume 5 • 1997

IMPROVING EDUCATIONAL PERFORMANCE:
LOCAL AND SYSTEMIC REFORMS

ADVANCES IN EDUCATIONAL ADMINISTRATION

IMPROVING EDUCATIONAL PERFORMANCE: LOCAL AND SYSTEMIC REFORMS

Editors: PAUL W. THURSTON
College of Education
University of Illinois at Urbana-Champaign

JAMES G. WARD
College of Education
University of Illinois at Urbana-Champaign

VOLUME 5 • 1997

Emerald

United Kingdom – North America – Japan
India – Malaysia – China

Emerald Group Publishing Limited
Howard House, Wagon Lane, Bingley BD16 1WA, UK

British Library Cataloguing in Publication Data
A catalogue record for this book is available from the British Library

ISBN: 978-0-7623-0081-5
ISSN: 1479-3660 (Series)

Transferred to digital printing 2006

CONTENTS

LIST OF CONTRIBUTORS

Robin L. Bird	Doctoral Student School of Education Stanford University
Robert D. Behn	Terry Sanford Institute of Public Policy Duke University
Sherry R. Eagle	Superintendent of Schools West Aurora, Illinois
Richard F. Elmore	Graduate School of Education Harvard University
Delwyn L. Harnisch	College of Education University of Illinois
Paul T. Hill	Institute for Public Policy University of Washington
Robert C. Hill	Superintendent of Schools Springfield, Illinois
Richard Kamm	Superintendent of Schools (Retired) West Chicago, Illinois
Michael W. Kirst	School of Education Stanford University
Arthur E. Lehr	Doctoral Student College of Education University of Illinois (UC)

Paul W. Thurston College of Education
University of Illinois (UC)

James G. Ward College of Education
University of Illinois (UC)

John F. Witte Department of Political Science
University of Wisconsin-Madison

DEDICATION

Edwin J. O'Leary
Teacher, Superintendent, Professor

IMPROVING EDUCATIONAL PERFORMANCE:

DIVERGENT POLICY PERSPECTIVES

Paul W. Thurston and James G. Ward

Education is a contested good in American society. There is little substantive consensus in American society about the goals of public education, the means to reach whatever goals do exist, and how schools should be held accountable. Since the publication of *A Nation At Risk* (National Commission on Excellence in Education, 1983), these and other related issues have been part of a national debate on how to improve educational performance and how to better meet the needs of an information age economy and society.

Educational policy makers disagree about the best approach to take to stimulate educational change and to improve student academic performance. Various approaches reflect value differences held by those influencing policy and by the world view of policy makers, key stakeholders in education, and the public. Some of these competing values and differing world views will be explored in this introductory paper. Subsequent papers will explore in depth many of these views and will reflect many of these value tensions.

Advances in Educational Administration, Volume 5, pages 1-12.

ISBN: 0-7623-0081-7

DIFFERING APPROACHES TO REFORM

Different approaches to changing schools are combined under the general heading of educational reform. There is, however, no consensus about the need for reform. Many critics of public schools, such as Bennett (1992) and Chubb and Moe (1990) decry what they see as the fundamental deficiencies of American public schools and call for radically different ways of educating American youth. They call essentially for the destruction of the American public school system as we know it because they feel it is unworkable and incapable of reform. Others, such as Berliner and Biddle (1995), Bracey (1996), and Rothstein (1993), argue that American public schools are doing much better than the public has been led to believe. Such different views of the current state of American public schools makes consensus on reform difficult, if not impossible. This explains, in part, the very different approaches that have been taken to school reform.

Labaree (1997) has identified five different perspectives on proposed education reform: the pedagogical, the organizational, the social, the cultural, and the political approaches. The pedagogical approach, he argues, is based on the assumption that the problem with American schools is that we have poor teachers and an inadequate curriculum. With this approach to reform, teacher preparation, recruitment, and evaluation and retention policies must be improved, and we must institute a much more rigorous curriculum. The organizational approach to reform argues that we must reform the educational bureaucracy to make it more responsive to desired educational outcomes, or we must eliminate the bureaucracy through private market approaches to education. Incentives are the key to improvement in the organizational view. In the social view of reform, schools must be improved by increasing attention to social problems such as poverty, racial discrimination, gender and social class bias, and social inequalities that produce unequal power situations. According to Labaree, the cultural approach to reform emphasizes confronting disintegrating family values, the culture of poverty, and the increasing gap between school culture and popular culture. Finally, Labaree argues that the central problem of reform is that we have not come to grips with the political problem of education, the conflicting and often unresolved multiple goals of schooling, and the inability of the system to satisfy all goals at the same time. This analysis would suggest that one way to read the papers in this book is to try to determine each author's view on the fundamental problem to be addressed, on why the public school system needs reform, and on how it can be accomplished.

One of our observations is that education reform efforts are often undertaken at the state level or in local school districts without a clear understanding of the nature of the problem being addressed or a clear conceptual understanding of the logic of reform. As a result, reform efforts are often disjointed and confused and have unclear purposes other than vague references to improved student performance. The papers in this volume are written to help clarify some of that confusion.

THREE OPERATIONAL FRAMEWORKS FOR EDUCATIONAL CHANGE

Three broad operational frameworks for educational change have been identified in education reform efforts to respond to the perceived failure of public schools to provide high levels of student academic performance: (a) local control, (b) state-driven reform, and (c) privatization. Those who advocate local control of schools emphasize the significance of the individual school building and the special relationships that exist between the school and the parents who send their children to the school. Those who advocate state-driven reform call for a more proactive role of the state in setting standards and in monitoring local school districts to be accountable in meeting those standards. Those who advocate privatization of public education call for breaking the government monopoly on providing education. Although the three approaches are compatible in many ways, each focuses on specific education components.

Local Control

Parents and other community members often ask, "What do good schools look like?" What attributes or characteristics are common to good schools? Sa (1993), an educational consultant who has visited many schools and thinks about school improvement, has created a list of common characteristics of good schools:

> a belief that all children can learn, although they may learn differently;
> an environment that is caring, personal, considerate and respectful of both children and adults;
> a sense of mission and vision that is shared by the entire school community;
> an organizing principle that places children's learning needs at the center of every activity, within a culture that values learning and insists on thoughtfulness, reflectiveness, self-assessment and self-examination;
> high expectations of everyone—adults and children;
> a knowledgeable staff that interacts regularly and frequently, collaborates to make decisions about all aspects of the school based on sound professional judgment and takes individual and collective responsibility for the success of students; and
> a curriculum—that is, a course of study that includes content, assessment practices, time, place and materials—that is coherent and based on rigorous inquiry (pp. 3-4).

Many school districts are trying to realize these good schools. They give attention and considerable autonomy to individual school sites. In this approach to school change, district administrators have service and support roles. School based management is an important component in such change initiatives within local districts.

School-Based Management

School-based management (SBM) is an organizational structure that is widely promoted. The precise implementation of SBM can differ considerably from district to district. At one level, SBM conveys greater autonomy at the school building level with less authority exerted at the district level. The governance system at the building level can differ widely. In many schools, the principal makes the decisions and the relationship between teachers and principal is no different than it was prior to SBM. In many other schools, SBM involves a more participatory governance relationship by teachers, one in which teachers actively participate in certain decisions, with a changed relationship between teachers and the principal.

SBM is a management approach intended to involve more people in the decision making process at the school level. Other words used to describe this approach include, school-based decision making, site-based decision making, and site-based management. For some, these terms carry significant differences in meaning, although for others, they are simply alternative terms to describe basically the same phenomenon. Hill and Bonan (1991), for example, distinguish SBM from shared decision making. SBM refers to the shift in decision making authority from the district to the school, whereas shared decision making describes the redistribution of authority among school site actors. This distinction recognizes the broad differences that exist from school to school as to who has authority to be involved in decision making at the building level. In many buildings, the principal retains considerable, if not all authority. In others, authority is assumed by teachers, parents, or elected citizens.

The conceptual basis for SBM rests on two basic assumptions. The first is that education will improve if those closest to the students have the discretion to make a variety of decisions about how they are educated and are held accountable for the students' academic performance. The second assumption is that the current problems in education are deep-seated, that major structural changes are required to remedy them, and that SBM provides the most compelling case for restructuring.

The assumption that greater discretion for those closest to the students will improve academic achievement is based upon several propositions. One proposition is dissatisfaction with the regulatory system characteristic of centralized control. This dissatisfaction is best described as a general agreement that some centralized regulations or requirements are inappropriate and that schools differ considerably from one another. Consequently, people working in these schools need the latitude to develop solutions that are most responsive to their particular needs. Moving authority closer to the students means providing more responsibility to parents, teachers, and principals. This responsibility will capture the creativity and energy of these people in ways that do not occur in centralized systems. The resulting sense of ownership and responsibility from this participation is more likely to produce effective and long lasting reforms in the schools than in the bureaucratic system (Stinnette, 1993).

State-Driven Reform

Dissatisfied with the performance of public education, many state legislatures and governors have taken a more directive approach to state educational policy since at least the early 1980s. Elmore (1991) distinguishes between the first wave of education reforms that took place from the late 1970s to the 1980s and the second wave that began in the 1990s. The first group of reforms, he explains, focused on academic content and intended to raise standards for teachers and students. The second focused "on fundamental changes in expectations for student learning, in the practice of teaching, and in the organization and management of public schools" (p. 1). Elmore explains the second wave of reforms as attempts to restructure schools and radically to change the nature of public schooling. The public, he notes, was skeptical that these reforms would be successful.

Throughout the 1990s the policy debate will continue over how much authority local districts should have in making educational policy decisions. O'Day and Smith (1991) and Smith and O'Day (1993) advocate systemic educational policy, also called content-driven systemic school reform, in which the state develops curricular frameworks and districts are free to apply them to their local contexts. Professional experts and citizens would provide input in the creation of frameworks that would "establish what students should know and be able to do" (Smith & O'Day, 1993, p. 251). All districts would be required to meet these standards. The framework would influence all state educational policies, including teacher certification, staff development, assessment systems, and curricular materials. At the local level, districts would have structured flexibility to implement the framework in their own way.

Numerous critics are concerned about the systemic approach (Clark & Astuto, 1994). Clune (1993), for example, outlines four problems with a mandatory system of state-directed, strong instructional guidance:

1. States are not in very good positions to develop frameworks as the systemic approach demands because of limited resources and the volatile political issues involved in such a task. The framework model also overlooks the possibility that some potentially effective reform initiatives may not lend themselves to being standardized to one model.

2. Delivering high quality education requires attention to a number of things in addition to high standards, the sole focus of the systemic approach. The state must also focus on getting real help to poor schools and obtaining real gains in student achievement.
3. The systemic approach emphasis on high standards will cause administrators to rely on high stakes examinations, decreasing educational quality by forcing teachers to teach to the test.
4. Most state governing structures are unable to implement a standard curriculum. Given budget constraints and historical top-down management, state educational agencies will not be able to implement the standard curriculum efficiently.

Despite these criticisms, Clune supports the basic premises of the systemic model.

Privatization

A variety of approaches have been implemented that utilize public tax revenue to support private delivery of education. Vouchers are being tested in Milwaukee and Cleveland with money going to parochial schools as well as private, nonreligious schools. The constitutionality of vouchers to parochial schools is being litigated. Charter school legislation, which authorizes the public funding of independently created schools, has recently been very popular. The criteria for chartering schools differs considerably from state to state. Contracting with private vendors to provide educational services also has recently been undertaken.

Increased privatization is supported by some to encourage competition, claiming that it will provide incentive for public school teachers to be less complacent about their current secure position. Through this competition, more efficient allocation of resources might also be attained. In addition, privatization is seen as a vehicle for encouraging the development of different models for delivering education.

FUNDAMENTAL VALUE ARGUMENTS AND ANTINOMIC ANALYSIS

Public policies are based on values held by those influencing policy and by the world view of policy makers and voters. As a result, the making of public policy on education has always felt the tension of competing values and differing world views. The usefulness of considering these contrasts in such an antinomic analysis is that it becomes clearer that each contrasting concept or principle has its strengths and its weaknesses. For example, a purely economic view can often miss important moral considerations. Political approaches may underestimate the importance of good administration (Carroll 1997). The following brief discussions highlight some of the antinomies that are reflected in the papers in this volume.

Decentralization Versus Centralization

Can schools best be improved through the creation of highly centralized governance and administrative systems that are tightly coupled and that can be easily controlled from some central point, or should the federal government or the states merely set clear goals and give local school authorities maximum discretion to make decisions about education? The United States has a strong tradition of local control of education and of considering semiautonomous states to be the "laboratories of democracy." By the same token, highly decentralized systems of schooling have produced racial segregation, huge disparities in per pupil expenditures among school districts, and unequal educational opportunities. In his discussion on systemic reform, Clune (1993) demonstrates some of the difficulties of trying to enact centralized reform in a decentralized system. Both centralization and decentralization have their merits, but we need to consider how a mixed system might work, if it can.

Politics Versus the Market

Can schools best be improved through better government decisions about the purposes and operation of schools, or should schools be left to a market environment where decisions are made by individual consumers of schooling? Advocates of the "common school" approach to education argue that schools need to be open to all and to be as all encompassing as possible in order to build a sense of community and to ensure equal educational opportunity. This can best be done through strong, government-operated schools that serve all children well. Market advocates argue that government is incapable of operating efficient and effective schools because of the influence of politics and that schools can only improve through open market competition fostered by educational vouchers, tuition tax credits, and other forms of finance promoting privatization of schooling. Charter schools have been seen by some as a way of combining these two approaches by selecting the best of the governmental approaches and mixing it with the best features of the market approach. This innovation is too new to have produced any solid results, but it merits consideration.

Professional Versus Community Control of Schools

Some people view schools as rational institutions with a clear technical core that need to be operated on the basis of professional expertise. Decisions about curriculum, instructional methods, student assessment, and the like should be made by professional educators who are steeped in the latest research and best practices and who can make the decisions that are the best for all children. Others claim that the problem with schools is that the community is shut out of school decision making and that public schools are estranged from their own communities because of professional control. They argue that schools will never improve until there is meaningful accountability of schools to the people and that the community can have direct influence over school professionals. These contrasting views both have to do with the legitimacy of public schools and their ability to build support among their constituencies. The tension between these two views is very evident in states where the state legislators have chosen to become micro-managers of the state school system.

Excellence Versus Equity

Some assert that the strong move toward equity in public schooling, which was dramatically advanced by the U.S. Supreme Court's 1954 decision in *Brown v. Board of Education,* intensified during President Lyndon B. Johnson's War on Poverty, and exemplified in the public school realm by the school finance reform movement, destroyed excellence and led to mediocrity in education. They claim that education is essentially meritocratic and that real excellence is not possible with a strong emphasis on equity. We can have equal educational opportunity, but we can never have equal educational results, they contend. Others argue just as strongly that an emphasis on excellence untempered by a strong concern for equity leads to elitism and undermines social justice. They assert that the problem with American schools is that we have focused on providing excellent schools for an elite few, often in affluent suburbs, and that the poor and less affluent have been left behind. For example, one argument for voucher programs is that they allow the poor in major cities to escape from pitiful inner city schools, as the affluent have always had opportunity to do. Another way in which this contrast has been articulated is through the question of whether we guarantee a high quality education for all children or whether we place our investment in education where it will have the greatest return.

None of these issues is simple and clear cut. Viewing the papers that follow in terms of antinomic analysis, however, can elicit insights and understandings that might not be as evident without such a viewpoint. These issues cut across the three parts of the book, and are often central to the positions taken in various papers.

OVERVIEW OF THE BOOK

The organization of the book tracks the frameworks for educational change described earlier in this paper. Part I proceeds from the perspective of local control, with special attention to measurement, motivation, and accountability. Part II focuses on curriculum content standards as a part of state-driven systemic reform. Part III flirts with the privatization perspective.

Robert Behn, in the paper titled "Linking Measurement and Motivation: A Challenge for Education," explores a central issue of local control, the relationship between measurement and motivation in education. After sketching the conceptual background for personnel accountability and the variety of reasons for measuring performance, Behn identifies a number of problems inherent in linking measurement to motivation. Despite these problems Behn argues that it is possible to design a system that links measurement to motivation and provides a number of hints about guiding the development of such a system.

Richard Elmore, in the paper titled "Accountability in Local School Districts: Learning to Do the Right Things," begins from the fundamental premise that the reason for local control of schools is that the local role can add value to student performance in schools. School districts operated within a performance-based accountability system where their success is measured by the ability of the individual schools to enhance student academic performance. Elmore is particularly interested in the capacity of districts to operate within this performance-based accountability system, what he calls "the problem of knowing how to do the right thing." Teaching practices need to be changed in order to increase student academic performance. One important aspect of changing teaching practices is to make existing knowledge about teaching and student learning accessible. Elmore provides some basic principles for designing systems that improve student performance. These include individual responsibility and system expectations. Responsibility, expectations, and formal accountability need to be aligned. Finally, school systems have a reciprocal duty to provide teachers with opportunities to acquire the knowledge and skills necessary to teach differently.

Robert Hill's paper, "Motivating Educators to Do the Right Things," provides a superintendent's perspective on ideas raised by Behn and Elmore. In his paper, titled "Democratic Accountability and Citizen Ownership: Political and Systemic Approaches to Improving Educational Performance," James Ward challenges fundamental notions of accountability in public education. Ward argues that considerable ambiguity exists about the things for which public schools should be accountable. Ward provides evidence that many people are satisfied with what schools provide and do not clamor for higher levels of student academic achievement. He also argues that more attention need be given to making better connections between the schools and the citizenry. Accountability must be to the people, according to Ward, not to narrow professional or elite group goals and purposes.

In Part II of the book, the paper by Michael Kirst and Robin Bird titled "The Politics of Developing and Sustaining Mathematics and Science Curriculum Content Standards," criticizes those who seek to avoid political conflict in setting curriculum standards. In reviewing the process whereby national math and science standards are determined, Kirst and Bird examine the political stresses brought about by conflicts among traditional, scientific, community, and individual interests in deciding what to teach. They also focus on problems in the way classroom teachers implement federal and state policies, especially when policy makers favor local control over central determination in setting curriculum expectations. Kirst and Bird note how attempts to bridge the gap between policy makers and educators have not succeeded in resolving tensions over participation in decision making, flexible versus prescriptive guidelines, and conservative opposition to more liberal education policies. Kirst and Bird recommend specific strategies to bring about systemic school reform, and assert that the key to change is linkage to broader political priorities.

In his paper, "A Classroom Perspective on the Politics of Curriculum Standards," Arthur Lehr suggests that consideration of the classroom situation should be added to the list of sources that influence curriculum standards. Although agreeing with much of what Kirst and Bird cite as important linkage between curriculum standards and classroom practice, Lehr supports Elmore's (1996) idea of policy makers sharing with classroom teachers the responsibility for meeting student performance expectations. Lehr argues for recognition of teachers as active participants in the process of developing curriculum expectations, noting that standards are not likely to be effective so long as those who prepare them work separately from those who apply them in the classroom. He also urges consideration of student motivation in setting performance standards, noting that the responses of students and teachers to the achievement expectations placed on them may depend largely upon how they see themselves as participants in determining those expectations.

Paul Hill and John Witte are well known for their writings about alternatives to public education: Hill for his appeals to more actively contract for private delivery of a variety of functions currently provided by public schools (Hill & Bonan, 1991; Hill & Byrk, 1996; Hill, Pierce, & Guthrie (1997), and Witte for his research on student performance in Milwaukee schools supported by vouchers (Witte, 1992, 1997; Witte & Archbald, 1985; Witte & Thorn, 1996; Witte & Walsh, 1990). Their papers, which lead off Part III of the book, deal with familiar themes, but put a different spin on them.

Witte, in "School Choice and Student Performance in Public Schools," directs his attention to the four major types of choice available in public education: magnet schools, interdistrict choice, open enrollment, and charter schools. He describes each of these alternatives and explains what is known about the effects they have on student achievement. Witte raises interesting questions about why so little attention is given to the phenomenon of choice in public schools while so much attention is paid to private school choice.

Hill, in "Accountability Under Charters and Other School Centered Reforms," focuses on the common theme of school centered reform in a variety of initiatives. School system decentralization, charter school laws, and public contracting all make the individual school a center of the reform effort. School centered reform initiatives share a common belief in the potential of the individual school site to efficiently utilize their resources and to effectively design their instruction. Hill is particularly interested in the accountability issues associated with school centered reform. He attempts to provide a conceptual framework for a school focused public accountability system where attention can be given to academic performance rather than bureaucratic measures. Papers by Eagle, Kamm, and Thurston provide critiques of the Hill and Witte papers.

REFERENCES

Bennett, W. L. (1992). *The devaluing of America: The fight for our culture and our children.* New York: Summit Books.

Berliner, D. C., & Biddle, B. J. (1995). *The manufactured crisis: Myths, fraud, and the attack on America's public school.* Reading, MA: Addison-Wesley.

Bracey, G. W. (1996). The sixth Bracey report on the condition of public education. *Phi Delta Kappan, 78* (2), 127-138.

Carroll, J. D. (1997). The warfare on and over American government in Waldonian perspective. *Public Administration Review, 57*, 200-204.

Chubb, J. E., & Moe, T. M. (1990). *Politics, markets, and America's schools.* Washington, DC: Brookings.

Clark, D. L., & Astuto, T. A. (1994). Redirecting reform: Challenges to popular assumptions about teachers and students. *Phi Delta Kappan, 75* (7), 513-520.

Clune, W. H. (1993). The best path to systemic educational policy: Standard/centralized or differentiated/decentralized? *Educational Evaluation and Policy Analysis, 15* (3), 233-254.

Elmore, R. (1991). *Restructuring schools: The next generation of educational reform.* San Francisco: Jossey Bass.

Elmore, R. F. (1996). *Accountability in local school districts: Learning to do the right things.* Paper prepared for the First Edwin O'Leary Symposium on Financial Management at the University of Illinois at Urbana-Champaign, October 16-17.

Hill, P., & Bonan, J. (1991). *Decentralization and accountability in public education.* Santa Monica, CA: Rand Corp.

Hill, P. T., & Bryk, A. (1996). *Toward a system of schools decentralization in six urban school districts.* Seattle: The University of Washington Graduate School of Public Affairs.

Hill, P. T., Pierce, L., & Guthrie, J. (1997). *Reinventing public education: How contract schools can improve American education.* Chicago: University of Chicago Press.

Labaree, D. F. (1997). Public education, private goods: The American struggle over educational goals. *American Educational Research Journal, 34* (1), 39-81.

National Commission on Excellence in Education. (1983). *A nation at risk: The imperative for educational reform.* Washington, DC: U.S. Department of Education.

O'Day, J. A., & Smith, M. S. (1991). Systemic school reform. In S. Fuhrman & B. Malen (Eds.), *The politics of curriculum and testing* (pp. 233-267). Bristol, PA: Falmer Press.

Rothstein, R. (1993, Spring). The myth of public school failure. *The American Prospect, 13*, 20-34.

Sa, S. (1993, Fall). We know what good schools look like, don't we? *NYRAG Times*, 1, 4-8.

Smith, M. S., & O'Day, J. A. (1993). Systemic Reform and Educational Opportunity. In S. Fuhrman (Ed.), *Designing coherent educational policy: Improving the system* (pp. 250-312). San Francisco: Jossey Bass.

Stinnette, L. J. (1993). *Decentralization: Why, how, and toward what ends*? (Policy Brief Report No. 1, pp. 1-5). North Central Regional Educational Laboratory.

Witte, J. F. (1992). Private versus public school achievement: Are there findings that should affect the educational choice debate? *Economics of Education Review, 11* (4), 371-394.

Witte, J. F. (1997). *Achievement effects of the Milwaukee voucher program.* Paper presented at the American Economic Association Annual Meeting, New Orleans, January 4-7.

Witte, J. F., & Archbald, D. (1985). *Metropolitan Milwaukee specialty schools and programs.* Milwaukee, WI: Report 5: Study Commission on the quality and Equity in the Metropolitan Milwaukee Public Schools.

Witte, J. F., & Thorn, C. A. (1996). Who chooses? Voucher and interdistrict choice programs in Milwaukee. *American Journal of Education, 104* (3), 186-217.

Witte, J. F., & Walsh, D. J. (1990). A systematic test of the effective schools model. *Educational Evaluation and Policy Analysis, 12* (2), 188-212.

PART I

REFORM AT THE LOCAL LEVEL: MEASUREMENT, MOTIVATION AND ACCOUNTABILITY

LINKING MEASUREMENT AND MOTIVATION:

A CHALLENGE FOR EDUCATION

Robert D. Behn

Two of the big questions of public management concern measurement and motivation:

> How can public managers measure the achievements of their agencies in ways that help to increase these achievements?
> How can public managers motivate people (public employees as well as those outside the formal authority of government) to work energetically and intelligently towards achieving public purposes? (Behn, 1995, p. 315).

These two questions apply to the management of regional public transportation systems, to the management of state departments of mental health, and to the management of municipal parks. They also apply to the management of public schools. We want to be able to measure the achievements of public schools in ways that help these schools achieve even more. We want to motivate teachers, principals, staff, students, parents, and citizens to achieve their schools' purposes.

Advances in Educational Administration, Volume 5, pages 15-58.

ISBN: 0-7623-0081-7

These two questions are connected; both focus on the challenge of achieving public purposes—on improving the performance of public institutions. Both measurement and motivation are of interest because, supposedly, they can help improve performance and, thus, increase the achievement of these purposes. The issues of measurement and motivation are directly linked, for one reason to measure performance is to motivate it. Presumably, if public managers employed both measurement and motivation well, they could improve the ability of public schools to educate children. Measurement and motivation have no value in themselves; they are only useful to the extent that they can help public managers improve performance.

HOLDING PEOPLE ACCOUNTABLE

Why, exactly, do we seek to measure the performance of public agencies? The most frequently offered answer is to hold people accountable. To hold them accountable for what? For whatever we are measuring, of course, and that might be anything.

We might want to hold public managers accountable for *inputs*—for having the proper inputs. In that case, we could measure whether each agency or unit hired people with the necessary or desirable credentials. What fraction of the teachers in a school had credentials that met the minimum standards? What fraction of the teachers had extra credentials? What fraction of the schools and of the teachers within the schools used the approved curriculum?

We might want to hold public managers accountable for *process*—for their diligent stewardship of the public's resources. We could, for example, measure whether each agency or unit paid its bills on time. In that case we could measure what percentage of an agency's bills were paid within, say, 30 days.[1] We could certainly measure whether an agency was operating within its annual, quarterly, and monthly budgets.

We might want to hold public managers accountable for *results*—for *outputs* or *outcomes*. The hold-people-accountable concept has come into vogue for this purpose. We want to hold public managers and public agencies accountable for results. We want to hold the regional environmental organization accountable for ensuring that the air in the region's airshed achieves certain clean air standards. We want to hold the welfare department accountable for finding jobs for welfare recipients and converting them into economically independent citizens. And, for example, the Dallas School Board wants "to hold all levels of the district accountable for results" (Clotfelter & Ladd, 1996, p. 29).

We might want to hold all schools accountable for educating their children. That is why we measure the performance of schools. As Michael Armacost, the president of The Brookings Institution, writes in the introduction of his organization's book, *Holding schools accountable*, "Many people believe that schools should be held more accountable for the academic performance of their students" (Ladd, 1996a, p. vii). "Systemic reform," writes Ladd, "requires that schools and school systems be held accountable for the progress of students toward the goals." Such reforms, observe Ladd and Clotfelter, "have sought to engender accountability by shifting the focus away from inputs to measured student outcomes" (p. 23). Ladd calls this "the new educational accountability" and "performance-based accountability" (1996b, pp. 4, 6, 13).

In 1987, the Government Accounting Standards Board (GASB) launched a research effort on the measurement of service efforts and accomplishments (SEA) "to determine whether the state of the art in SEA measurement is sufficiently developed to warrant" the inclusion of SEA indicators in the financial reports of state and local governments (Hatry, Alexander, & Fountain, 1989, p. 4). "Such financial reports should," stated the GASB, "provide information to assist users in (a) assessing accountability and (b) making economic, social, and political decisions." In creating its SEA project, "The Board gave considerable weight to the concept of accountability: of 'being obliged to explain one's actions, to justify what one does'; of being required 'to answer to the citizenry—to justify the raising of public resources and the purpose for which they are used'" (Hatry, Fountain et al., 1989, p. v).

What is this concept of accountability? What does holding people accountable really mean? Beyond explaining one's actions and justifying the use of public resources, what does it mean to answer to the citizenry? When we hold someone or some organization accountable for something, what do we really do? In some ways, it means that we want to be able to identify who is responsible for the organization's outputs or outcomes, for its successes and failures. But then what? That answer does not really clarify things. What does it mean to hold people responsible for success or for failure?

I know of no definitive answer, either theoretical or empirical. But I bet I know what the managers who are to be held accountable think. I bet they believe, from their own empirical experience, that holding people accountable means that when they fail they are punished and that when they succeed nothing significant happens.

Hanushek and his colleagues observe that "state and federal governments, like local schools, exhibit a distinct lack of enthusiasm for evaluating schools" (1994, p. 145). Yet, most public managers realize that any new system designed to hold them accountable for results means, operationally, an increase in the probability that they will be punished for some kind of failure and little change in the probability that they will be rewarded for any kind of success. Should we really be surprised or puzzled that people resist being held accountable?

Now certainly, programs designed to hold schools accountable for results usually reward a certain fraction of the schools for their performance. These rewards, however, are usually limited in size and usually limited to well less than half of the schools. From the beginning, therefore, such holding-people-accountable systems are predestined to label as failures more than half of the school districts, more than half of the school superintendents, more than half of the schools, more than half of the principals, and more than half of the teachers. Then why are we surprised that many of the people and organizations we seek to hold accountable are resistant to the entire concept?[2] How attractive is the opportunity to be punished even more?

Usually, holding-people-accountable systems are supposed to reward successful organizations and to help the failures. Schools ranking at the bottom in terms of performance may receive extra attention, guidance, or even resources. Armacost writes of "accountability and incentive programs that reward the most effective schools and generate increased scrutiny of the least effective ones" (Ladd, 1996a, p. vii). There is no punishment here, just rewards for the top schools and opportunity for improvement for those at the bottom. Hanushek and his colleagues, however, are more straightforward: "Performance incentives are rewards and punishments related to the achievement of specific outcomes" (1994, pp. xx-xxi).

Failing to reward a school is, operationally, the same as punishing it. The school did not get something, and that something is the reward. It is like the bonus question on a mid-term exam. The teacher explains that you are not penalized for not doing this question, but that you can earn extra points if you answer it correctly. The clever student quickly figures out that, when the points are totaled, the extra bonus points count just like the regular points. Failing to tackle and earn points on the bonus question has precisely the same consequence as failing to tackle and earn points on a regular question. A holding-people-accountable system that rewards success by distributing some kind of bonus or prize must, whether it seeks to or not, punish failure by withholding that prize.

Rewarding high-scoring schools with increased flexibility has the same consequence for the low-scoring schools as does rewarding the top schools with more money. The effect is to punish the low-scoring schools with (comparatively) more regulations. In the future, in addition to whatever problems caused these schools' low performance, they will have some new, additional hurdles to overcome: They must compete with the higher-performing schools while burdened by the internal demoralization that came from the published scores, they must compete with schools that now have extra resources, and they must compete with schools that now have more flexibility.[3] Do we really think that such a holding-schools-accountable system is the best way to help these low-performing schools to improve?

FIVE REASONS TO MEASURE PERFORMANCE

"Performance measurement," write Blodgett and Newfarmer, "is arguably the hottest topic in government today" (1996, p. 6). But why? Why should public managers develop measures of their organizations' performance? Public managers have a number of different tasks to perform. Can performance measures help them carry out any of these responsibilities? Pillsbury (1996) has suggested that performance measures have five different uses: Managers can measure performance to help them: (a) evaluate, (b) motivate, (c) learn, (d) promote, and (e) celebrate.

Evaluate

A manager needs to be able to evaluate units and people. Measures of performance help the manager do this by providing a basis of comparison. If the measures are selected well, they can be used to determine which units and people are doing a good job and which are not. Comparative evaluation is the traditional, primary and, often, sole reason for measuring performance.

How can we determine which schools are educating children and which ones are not? The answer is: measure them. Create some kind of measure, some kind of test, to determine how much each student knows. We implicitly attribute what a school's students know to the performance of that school. Such evaluation is the fundamental rationale behind the "report cards" that many state education departments issue giving the test scores for their school districts.

Motivate

A manager also needs to motivate people. If people know that they will be evaluated on the basis of particular measures, they will be motivated to achieve better measures. Even if the measures are not linked in any formal or even informal way to an explicit evaluation system, the mere publication of comparative performance measures can motivate behavior. People naturally want to look good, particularly to their peers. Thus, they will seek to look better in terms of published measures, be they home runs hit, tax dollars collected, or mathematical concepts mastered by students.

Measures are assumed to motivate performance. Education measures are assumed to motivate educational performance. This is a second purpose behind a report card for a state's school districts. If we publish the test scores of districts and of schools, we will, presumably, motivate their principals and teachers to concentrate on improving these test scores.

Learn

Organizations need to learn. But how can the leaders and members of an organization learn if they lack feedback on their progress? Although the leaders of a public agency might want to use one performance measure to motivate behavior, they need to collect a lot of other measures of performance to learn.

By collecting a wide variety of data, the leadership can not only discover how the measures are motivating and distorting behavior; they can also learn what units are doing better than others and use this knowledge to help the rest of the organization improve. By collecting and analyzing lots of data, they can learn how they need to change the measures they are using to better motivate behavior.

Measures can help a school district to learn on what specific topics each school needs to concentrate to improve. Measures can also help a district learn which schools and teachers are doing a good job and then employ these people to assist those schools that most need improvement.[4]

Promote

Any organization and, particularly today, any public agency needs support. It needs to establish that it is doing a good job. Only by demonstrating that it is accomplishing important purposes can the leaders of a public agency create the support they need to continue to achieve these purposes and, perhaps, accomplish even more. By publicizing an agency's performance measures—how it is doing, how much progress it has made, how well it has done in achieving its goals—the agency's leaders can build the support necessary to generate resources to ensure continued performance.

Schools and districts that score well or improve their scores significantly on widely-accepted performance measures can use this success to dramatize that the taxpayers' resources have been well spent and to argue that such funding should be continued. Schools and districts that score poorly can use performance measures to argue that more resources are needed to improve performance. All schools can use performance measures to promote their past accomplishments and to explain exactly what additional resources can produce in terms of even better, future accomplishments.

Celebrate

Organizations not only need to accomplish purposes. They need to celebrate these accomplishments. By setting particular targets and then measuring progress toward achievement of these targets, public agencies can create a sense of accomplishment for those within the agency. By publicly celebrating the periodic realization of specific targets, the agency's leaders can build internal and external morale. Further, such a celebration can promote the agency outside and motivate further progress inside. The opportunity to celebrate success is valuable only to the extent that it helps motivate or promote.[5]

The publication of a school's annual performance measurement in the state's report card can be but one opportunity for a school to celebrate. Such a celebration rewards past accomplishments, motivates future efforts, and promotes the need for sustaining or increasing the level of resources.

What reasons do others offer for measuring performance? The GASB has undertaken to move beyond the standard financial measures to create measures of service efforts and accomplishments (SEA) because it wants its data to make a better contribution to the performance of government. How might such SEA data be used? To supplement the generic concept of accountability, the GASB offered several reasons for measuring SEA:

> SEA information is needed for setting goals and objectives, planning program activities to accomplish these goals and objectives, allocating resources to these programs, monitoring and evaluating the results to determine if they are making progress in achieving the established goals and objectives, and modifying program plans to enhance performance (Hatry, Fountain et al., 1989, p. v).

One of the key issues raised by GASB's research concerned the uses and users of SEA data. GASB concluded that SEA measures can:

1. ... provide much *greater accountability* of governmental entities...
2. ... be used as a *major basis for motivating public employees*...
3. ... *stimulate the public to take greater interest in governmental entities and provide more encouragement* to officials to provide quality services.
4. ... help *determine and justify the need for governmental programs*...
5. ... [create a situation in which] public policy discussions may be more likely to *focus on issues concerned with program results and have a more factual basis*].
6. ... *encourage improvement in government programs and policies* (pp. 33-34).

When moving from SEA reporting in general to SEA reporting for elementary and secondary education,[6] GASB identified five basic uses for such data:

1. To achieve accountability by informing the public and their representatives of the result of operations.
2. To improve educational programs.
3. To encourage public interest and pressure for improvements.
4. To provide a basis for incentives and sanctions.
5. To monitor the cost associated with a particular amount of improvement in performance (Hatry, Alexander, & Fountain, 1989, p. 71).

None of these GASB statements mention using SEA data to celebrate success. Nevertheless, once you eliminate the vague, nonoperational notion of achieving accountability, GASB's remaining purposes match up quite well with Pillsbury's four remaining reasons for measuring performance: to evaluate, learn, promote, or motivate.

LINKING MEASURES TO MANAGERIAL TASKS

Why do public managers measure the performance of their organizations? To help them with their basic, managerial tasks. I am not seeking to create the definitive list of purposes for measuring performance, but I do want to emphasize the necessity of being explicit about such purposes. I want to emphasize the importance of thinking through the link between the measures being collected and how they will be used by public managers to undertake their basic tasks. Holding people accountable is simply too ambiguous a concept; it provides little guidance for designing and employing performance measures. A more explicit statement of purposes is required. Until you understand how you will use various measures of performance, it is difficult to decide what to measure.

The leaders of a public agency ought not to create performance measures in the abstract. Rather, they should choose carefully; they should design each measure to help them with a specific managerial chore.

Furthermore, there is no reason to assume that one measure will satisfy all of the manager's needs. For example, the measure used to motivate behavior inside an organization might be quite different from the measures used to promote the organization with outside stakeholders and citizens.

The choice of measures to use for different purposes will depend on an organization's history and status. To motivate people inside an agency, for example, its leaders should choose measures that reflect where the agency is at the moment, what challenges it faces, what problems it has, where it needs most to improve, and what it needs to accomplish next. Then, the leaders need to continually evaluate the measures they have chosen and modify them as needs and circumstances change. The measure that will help one school motivate the performance improvement that it needs to make may not be the measure that another, different school needs to motivate the achievement of its own, different performance challenge.

We should not assume that one single measure is necessarily the best for all managerial purposes. Neither should we assume that the measure for one purpose is the right one for all similar organizations. Nor should we assume that the measure that is best for one of an organization's purposes today will be best for that same purpose next year. There is simply no measure for all seasons.

THE THEORY OF THE MEASUREMENT-MOTIVATION LINK

How is measurement linked to motivation? The most b asic answer comes from Haire's observation: "What gets measured gets done" (Peters & Waterman, 1983, p. 268). To illustrate, Osborne and Gaebler quote a former Massachusetts commissioner of public welfare whose department was plagued by a high error rate:

> When we began measuring error rates statewide, nothing happened. But as soon as we published the rates for each office, things changed. Because now the finger was pointed

> at the managers; everyone knew if their office had a high rate. It only took 12 months to cut it down [from 23] to 12 percent. Six months later, when we published error rates for each supervisor, it fell to eight percent. As this commissioner explained: "All you have to do is measure something and people respond" (Osborne & Gaebler, 1992, p. 146).

Why do people respond to measures? "What gets measured gets done" is simply an empirical observation without any connecting theory. Why would we do something simply because it gets measured? The answer comes from Chase's observation: "Nobody likes to look bad in a crowd" (Chase, 1979, p. 433). Measurements create comparisons. If you measure something and publish the results, you implicitly compare people. Even if you do not order the listing by how people scored on the measure, readers will quickly create the relevant ranking in their heads. People will identify whom they want to compare with whom and check their relative scores to see who is better. The mere publication of the results—without an ordered ranking or any accompanying commentary—clearly identifies who is doing well and who is doing poorly.[7]

Chase, however, took the measurement-motivational link one step further by tying the measurement to a reward in what has come to be known as "Chase's Law":

> Whenever the product of a public organization has not been monitored in a way that ties performance to reward, the introduction of an effective monitoring system will yield a 50 percent improvement in the product in the short run (Allison, 1980, p. 38).

This, of course, makes the motivational link more explicit. What gets done is not what you measure but what you measure and reward. The reward, extrinsic or intrinsic, provides the link between measurement and motivation.[8]

Thus, the typical efforts to hold public organizations accountable for performance—and to hold public schools accountable for performance—have been built around some kind of measurement system linked to some kind of reward (Ladd, 1996b). The measurements are designed to capture school performance by determining how much students learned, either by testing the students or evaluating portfolios of their work. Then those schools that do particularly well receive rewards; such rewards can either be money (bonuses to districts, schools, or teachers), favorable comparisons (created by the publication and ranking of all schools' scores), symbols (such as flags that a high-performing school can fly), or increased managerial flexibility (the suspension of some overhead rules). In contrast, those schools that do poorly are punished; they may lose funds or be subjected to closer administrative oversight. The poor performers, of course, are also punished through unfavorable comparison, for everyone knows that they ranked at the bottom. What school superintendent wants to go to the annual state meeting just after the state's report card has been released when his or her school has been ranked near the bottom?

Money and Motivation

What is the best kind of reward? In a market economy, the answer is money. More money is, by definition, an indicator of success. Less money is an indicator of failure. So, how better to reward people than to give them more money. In fact, we often implicitly assume that any real reward has to mean money.[9] More money is better than less money. Everyone would like more money; pure money is fungible. Why reward people with a snorkeling trip to the Bahamas if they would really prefer to go skiing in Idaho or to extend their collection of antique baseball cards. Money is such an obvious reward that it hardly needs to be justified or analyzed.

Money, however, may not be the best motivator, at least for individuals. Herzberg (1968) has argued that, although money handled badly can be a significant demotivator, money handled well is rarely a motivator. This is because, concluded Herzberg, "The opposite of job satisfaction is not job dissatisfaction but, rather, no job satisfaction; and, similarly, the opposite of job dissatisfaction is not job satisfaction, but no job dissatisfaction" (p. 56). Thus, Herzberg distinguished between the impact of two different sets of job characteristics: (a) hygiene factors that, if handled badly, produce job dissatisfaction but, if handled well, merely result in no dissatisfaction; and (b) motivators that, if handled well, do produce job satisfaction and motivation but that, if handled poorly, merely result in no job satisfaction. Significantly, salary was one of the hygiene factors that could only, if handled badly, produce dissatisfaction; salary rarely produces satisfaction. Herzberg offered a list of motivators: achievement, recognition, the work itself, responsibility, advancement, and growth.

For elementary and secondary education, that looks like good news. After all it is, or ought to be, relatively easy to give teachers an opportunity for achievement, recognition, responsibility, advancement, and growth. The inherent characteristics of the teaching profession and the nature of the work itself ought to facilitate the creative use of these motivators.

In contrast, giving teachers more money will rarely motivate them. But handling salary badly can create extreme dissatisfaction. Given some of the inherent characteristics of American government—political instability, competing interests and theories, impulsive decision making—public managers may never be in a position to distribute bonuses and merit raises with the care, consistency, and subtlety necessary to have them be anything but a demotivator. Consequently, public managers might attempt to eliminate or neutralize the negative impacts of Herzberg's hygiene factors by reducing the importance of money and minimizing its negative impact. They would be more effective by focusing their motivational efforts on Herzberg's "motivator factors."[10]

There is evidence to suggest that, from the menu of potential motivators available to public managers, money may not be the most effective. Over half a century ago, Maslow (1943) defined a hierarchy of five needs that motivate human behavior: physiological, safety, love, esteem, and self-actualization. Once an individual satisfies one need, he or she moves up the hierarchy and is motivated by the desire to achieve the next need. For example, Maslow writes that "if both the physiological and safety needs are fairly well gratified, then there will emerge the love and affection and belongingness needs" (p. 380).[11]

Maslow's need for love includes not only romantic love but also friendship, a person's need for "a place in his group." The next need, esteem, includes both self-respect and the esteem of friends and peers; it is "soundly based upon real capacity, achievement, and respect from others" and includes "the desire for strength, for achievement, for adequacy, for confidence in the face of the world, for independence and freedom" as well as "the desire for reputation or prestige (defining it as respect from other people), recognition, attention, importance, or appreciation" (pp. 381-382). Certainly, these esteem needs could motivate someone to choose teaching as a profession, and, moreover, the profession could fulfill them. Although money can directly purchase the first two needs (physiological and safety), it can only *help* in obtaining love and esteem.[12]

Self-actualization, the fifth and highest need in Maslow's hierarchy, is the "desire for self-fulfillment...to become everything that one is capable of becoming" (p. 382). Maslow notes that "in one individual it may take the form of the desire to be an ideal mother, in another it may be expressed athletically, and in still another it may be expressed in painting pictures or in inventions" (p. 383). Certainly, self-actualization can come from being an exceptional teacher.[13] Except to the extent that money can give someone the complete freedom to pursue self-actualization, it is irrelevant to the fulfillment of this need.

Consequently, if an organization's basic salary levels are not high enough to satisfy its employees' physiological and safety needs, money will be a powerful motivator. But what happens when salary levels are adequate to cover these two basic needs for most of the organization's employees? What if an employee's wealth and income are sufficient to purchase the level of food, housing, and security to which he or she would like to be accustomed? Then, money may not be the best motivator.

Public sector employees may switch from physiological and security needs to a focus on love and esteem at lower levels of income than other people. People who choose to work for government agencies may be precisely those whose physiological and security needs are lower than average. People who seek employment in an environmental agency are not motivated by the lure of earning enough to buy a 65-foot yacht; these people want to save the planet. People are not motivated to become teachers by the size of their office or the perks of corporate travel; these people think, initially at least, that they will enjoy the love and esteem that comes from being a good teacher. They might even believe, again initially, that a teacher's salary and social status satisfies their first four needs, leaving them free to pursue self-actualization through teaching.

There exists enough general information about the salaries paid by government and the purposes it pursues to encourage self-selection. People motivated more by physiological and security needs than by the love and esteem that comes from achieving public purposes will seek employment that pays enough to satisfy these needs at whatever level they think appropriate. People whose physiological and security needs can be satisfied at lower levels[14] and who seek the love and esteem that comes from achieving public purposes are more apt to seek employment with a public or nonprofit organization, particularly with an organization that is pursuing purposes to which they would like to contribute.

Public agencies ought to strive to benefit from such advantageous selection.[15] The U.S. Army does not recruit people by emphasizing its salaries or even its food. Rather, it advertises that the army gives the individual a chance to "be all that you can be." Everyone knows that you will not get rich by joining the military. Thus, the U.S. Army recruits people by promising an opportunity at self-actualization.

THE MEASUREMENT-MOTIVATION LINK IN EDUCATION

Education is a data-rich policy area. We frequently test students, and then we collect, analyze, and publish data on these tests. Consequently, it is easy to compare such data on anything from individual students to nations. Such data provide an attractive and easily available lever for holding teachers, schools, and districts accountable. Collect the data (being careful to do it right to eliminate the various and obvious sources of bias); publish the data; reward the successful teachers, schools, and districts; and punish the unsuccessful ones. Then, according to the assumption behind the holding-people-accountable theory, things will automatically get better.

But why? What are the mechanisms that link educational measurement to educational motivation and, eventually, to educational improvement? Is it the allure of the reward and the fear of the punishment? Is this the motivational link? If so, we ought to examine this proposed linkage more carefully. We ought to understand exactly how it works and under what circumstances it works. We ought to think seriously about how the rewards and punishments of any holding-people-accountable system can or cannot link measurement to motivation.

Two kinds of rewards are deployed to hold people accountable: extrinsic or financial rewards and intrinsic or psychic rewards. We can provide people with extrinsic rewards, which can best help them satisfy their physiological or security needs. Teachers, schools, and districts are awarded with more money if they succeed and perhaps lose money if they fail. Alternatively, we can provide people with intrinsic rewards, which can help them satisfy their love or esteem needs. Teachers, schools, and districts are recognized publicly for their accomplishments and successes and recognized publicly for their failures, too.

What kinds of rewards and punishments are most effective in linking measurement to motivation? How big must a reward or punishment be to motivate? If neither the rewards nor the punishments are very large, they may have little motivational impact. How big are the rewards and punishments in comparison with the effort necessary to win the rewards or to avoid the punishment? How big are the rewards and punishments in comparison with other rewards and punishments that are generated by other, perhaps conflicting, behavior? At a minimum, doing nothing has an obvious reward: more time available for other, more rewarding activities.

Doing nothing avoids the built-in psychological punishment that comes from trying and failing. If you are looking for self-esteem or the esteem of others, if that is where you are on Maslow's hierarchy of needs, taking on a project doomed to failure hardly makes sense. It can only undercut your own self-esteem and, if your failure is publicized, the esteem of others.

All this suggests that, given the challenge of improving American elementary and secondary education, particularly the challenge of improving American education where it most needs improvement, the rewards need to be large. But will such large rewards ever be forthcoming from state legislatures? Or, more importantly for motivational purposes, will teachers, principals, and superintendents ever believe that large, promised rewards will ever be delivered? And, if they are miraculously delivered once, will these people believe that large levels of rewards will be repeated legislative session after legislative session? Do we want to commit ourselves to a holding-people-accountable strategy that depends upon an uninterrupted stream of large, financial rewards?

Are financial rewards the best kind for connecting educational measurement to the motivation of teachers? What is the motivational impact of giving teachers bonuses, large or small, for producing better results? Will financial rewards tap into the needs that motivate teachers? There is evidence that such bonuses may be counterproductive. By paying more for dedicated work, public agencies may undercut their employees' higher motivations. Cialdini (1993) summarizes some of the relevant research on motivation:

> Social scientists have determined that we accept inner responsibility for a behavior when we think we have chosen to perform it in the absence of strong outside pressures. A large reward is one such external pressure. It may get us to perform a certain action, but it won't get us to accept inner responsibility for the act. Consequently, we won't feel committed to it (p. 93).

Most government agencies pursue public purposes that are very difficult to achieve. Thus, to produce real results, these agencies need employees who will make a personal commitment to achieving these results. A large reward, particularly a monetary bonus that is transparently a reward, may undercut that commitment.[16]

Teaching requires personal commitment. It is not assembly line work. It is a job shop. Teaching each student is a separate task requiring its own unique strategy. To teach each student each subject may be its own separate task. Devising and implementing all these individual strategies requires a lot of effort, a lot of time, and a lot of commitment. Fortunately, the personal choice of making that commitment may be highly motivational.

But to what is a teacher committed? To teaching well or to making money? If the teacher is teaching well to accomplish a social purpose, the commitment makes sense. But if the teacher is teaching well to make more money, the commitment is silly. Why teach to make more money? Why not simply get into a more lucrative profession? If more money is the objective, changing careers is a much more effective strategy than teaching better. Using financial rewards to motivate teachers, particularly given the small size and uncertainty of any such reward, may actually undercut the teacher's personal commitment to teaching well. If nothing else, the bonus reminds the teacher of how meagerly and begrudgingly our society rewards excellent teachers.

Raising the salaries of all teachers will, over the long run, expand the pool of people willing to enter the profession. If we all need to satisfy our physiological and safety needs before we can begin worrying about love and esteem and if we each have a different level of these two basic needs, raising the salaries of all teachers will give more people an opportunity to consider obtaining their need for love and esteem through teaching. More money may motivate more people to consider a career in teaching, but once people have chosen this profession, more money—particularly the small amount that legislatures may allocate to teacher bonuses—is unlikely to motivate them to work significantly harder.

A bonus awarded publicly for a significant accomplishment has motivational value beyond its monetary value. The real value of such a financial reward may lie in what it symbolizes: a significant contribution to the learning of many students. The recipient of the bonus award may well have understood the significance of his or her contribution, but by bestowing the award publicly, the organization enhances the recipient's self-esteem and esteem from peers, friends, and the general public.

If the real value of the financial reward is what it symbolizes—if the motivational impact of the award comes from its contribution to an individual's sense of belongingness and esteem—why trivialize the symbol? Why dilute its motivational impact by coupling it with a pittance of money?[17] Why not make the symbol big and public and ignore the money? Why not focus on the recipient's clear need for love, affection, belongingness, and esteem? Why not explicitly recognize and reinforce the recipient's personal commitment to teaching? Why even suggest that the recipient did it for the money? Why use an ineffective motivator when a more potent one is quite available and does not require legislative approval?

TEN PROBLEMS OF LINKING MEASUREMENT TO MOTIVATION

Anyone who deploys measurements to motivate discovers that it is not all that easy. For despite Haire's assertion that what gets measured gets done, what actually gets motivated, let alone done, may not be precisely what those doing the measuring intended. Using measures to motivate creates a number of practical problems.

Establishing Measures is Controversial

Measures are designed to capture how well agencies are achieving specific public purposes. Because many public purposes are controversial, however, establishing a method of measuring the accomplishment of these purposes can be similarly controversial. The lofty, vague words with which public purposes are often expressed can disguise fundamental, underlying disagreements. Only when we explicitly consider how to measure the achievement of these purposes may these conflicts become obvious. Thus, debates about measures are often surrogate debates about purposes. Consequently, to avoid the conflict that any such debate will generate, to avoid exposing the underlying disagreement about purposes, people eschew even attempting to select specific performance measures.

The controversial nature of selecting performance measures certainly applies to education. "Consensus on the overall goals of American education has, at least in the past, been elusive," note Hanushek and his colleagues. Consequently, "The question 'what to measure?' is one of the most contentious issues in standardized testing" of students (1994, p. 133).

It is Often Impossible to Measure Outcomes

Every public agency has been created to produce some kind of improvement in society, some kind of result, some kind of outcome. The real outcome that we seek from a public agency, however, may not be realized or measurable for years. If we want some kind of useful measure of how well the agency is doing now, we have to rely on its outputs or some other outcome surrogate—some other that demonstrate some measure of what it has done this year.

We want our schools to fill students' heads with knowledge and analytical skills. Yet, we want our schools to educate students about more than the standard textbook subjects from reading to physics. We want our schools to do more than teach students to think analytically and creatively. We also want our schools to create future contributors to society. We want schools to help children mature into economically productive and socially responsible citizens.[18] This is the ultimate outcome that we really want our schools to produce. This final result is certainly difficult to measure; moreover, it is very difficult to measure this result in any manner that can be used in any reasonable way to help improve a school's current performance.

The real outcomes of many public agencies do not appear for years or decades. Rather than measure the future outcomes we really want to achieve, we are forced to measure the outputs that we are actually producing now: the knowledge and analytical skills, for example, that are inside the heads of today's students. We exercise the natural human tendency to measure that which we can measure best, such as current students' ability to solve specific problems.

Measures Can Distort Behavior

Not only do measures motivate behavior in desirable ways, they can also motivate behavior in undesirable directions. As Hanushek and his colleagues observe, "The problem with applying performance incentives to the management of education is defining what sort of incentives will work most effectively without also having undesirable side effects" (1994, p. 55). Similarly, Ladd worries about "whether the undesirable side effects of accountability and incentive systems can be kept to a tolerable level" (1996b, p. 14).

In part, this problem emerges from our inherent inability to measure precisely whether we are achieving our public purposes. Usually, an actual measure is only a crude approximation of what we want to accomplish. For example, we want to educate children to be able to use mathematical thinking to solve a variety of complex and new problems. Yet, standardized tests usually ask students to use mathematical formulas to solve a number of standard problems. Students who do better on such problems will also do better when confronted with more complex problems. After all, understanding how to do simple mathematical problems is a prerequisite for undertaking more complicated analytical reasoning. Nevertheless, what we measure is not precisely what we want; it may not even be close.[19]

As Haire (Peters & Waterman, 1983) reminds us, whenever we measure something in education, or any business, people will respond. Measurement can produce that fortunate and unfortunate result: exactly what we asked for. But, because we cannot measure what we want, what we get is what we asked for, not what we wanted. We suffer from our human limitations, our inability to measure the actual achievement of our real purpose. What gets done, therefore, is not what we really wanted but only what we could measure.

Of course, the people being measured are not stupid. They know what we really want. Nevertheless, they do what is being measured, either out of narrow self-interest (they want the reward that comes from producing the desirable measurement) or out of contrariness (they think it is stupid but decide that, if that's what is wanted, that's what they will do). The result may be cheating, either honest cheating or dishonest cheating.[20] The distinction between honest cheating and dishonest cheating is similar to George Washington Plunkitt's distinction between "honest graft" and "dishonest graft": honest graft is quite legal while dishonest graft is not (Riordon, 1963, pp. 3-6).[21]

Honest cheating is knowingly doing precisely what will be measured rather than seeking to accomplish the real purpose behind the measurement. In education, the obvious form of honest cheating is teaching to the test. The real purpose is to teach children to think mathematically. Because it is quite difficult, however, to measure our success at achieving this purpose directly and reliably, we resort to a surrogate: a standard test. Because this limitation is well known, because all teachers understand that it is impossible to measure the level of achievement of our true purpose, they can focus on the measure being used, not on the real purpose. They need fear no retribution. Yet, it is hard to fault people for teaching to the test or engaging in other forms of honest cheating. After all, they were told quite clearly how their performance would be measured.[22] The teachers were not told to achieve the real, but vague, purpose. They were asked to produce against a proxy, but specific, measure. As George Washington Plunkitt would often say: "I seen my opportunities and I took 'em" (Riordon, 1963, p. 3).

In contrast, dishonest cheating is, well, dishonest. It is clearly unethical if not downright illegal. The obvious example is the elementary school in Fairfield, Connecticut, that in 1996 was accused of erasing its students' erroneous answers and replacing them with correct ones. This is certainly dishonest cheating.

All measurements can distort behavior. They encourage people to focus on the measurement rather than the real purpose. They inevitably encourage both honest and dishonest cheating. And, when the measurement is linked to some kind of reward, when the stakes of doing well against the measure are raised, the incentive to cheat does increase.

Traditional Holding-People-Accountable Systems Create Many More Losers Than Winners

What happens once the results are in, once all the measures have been recorded and tabulated? Usually, in a system designed to hold people accountable, some people or organizations receive rewards. The rest receive nothing or, perhaps, even some kind of punishment. Moreover, the number of awards is limited. Before the measurements were ever taken, the number of awards was arbitrarily fixed.[23] South Carolina, for example, gives awards in the range of $15,000 to $20,000 plus some managerial flexibility to the top

25 percent of its schools. Dallas fixes not the number of schools receiving its awards but the amount of money to be distributed to both the winning schools and their teachers and other staff; approximately 20 percent of the employees in the Dallas schools receive awards each year[24] (Clotfelter & Ladd, 1996, p. 35). For both these award programs, approximately three-quarters of the schools and three-quarters of the principals and teachers are labeled losers.[25] Is this motivating?

If your school just missed the cutoff this year, you and your colleagues might well be motivated to do better next year—to work hard enough to win one of the awards. (If your baseball team just missed the playoffs this year, you and your teammates might well be motivated to do better next year—to work hard enough to make the playoffs.) But what if your school finished near the bottom? What if the improvement required to become one of the winners is colossal? Will your school really be motivated by this kind of reward system to strive to become one of those top twenty-five percent schools? How hard will you and your fellow teachers try if you know that you not only have to improve but also leap over at least half of the other schools to make the top twenty-five?[26] (How hard will you and your baseball teammates work if, after two months, you find yourself at the bottom of your league? How motivated to make the playoffs will you be once you realize that to do so your team must not only perform spectacularly but also all the other contenders have to collapse?)

Clotfelter and Ladd observe that "most recognition and reward programs perform some kind of adjustment of the absolute scores in an effort to give all participants a reasonable chance to be a winner" (1996, p. 26). This kind of adjustment is designed to ensure that the measurements do not discriminate against poor or minority districts or schools—to ensure that they have a chance to win. But if the number of rewards is fixed, not every district or school can win. All such an adjustment is doing is trying to make sure that every school has the same one in four chance of winning.

We Americans love competition. We believe in competition. We know that competition motivates. Sports is our model for competition. All the teams in a league compete to be, at the end of the season, the single, lone champion. Such competition is motivating for those who have a chance at winning.

Moreover, the competition is exciting. It is dramatic. It is fun to watch as teams drop out, as fewer and fewer teams remain and, in the end, one team is crowned the champion. It makes for great entertainment.

We are not, however, naive. We know that if we played out the season again, the result would not be identical. Maybe the Chicago Bulls would win again—just as did the Boston Celtics and the New York Yankees of old. But the result would never be identical. Different teams would win some different games; often the same team would win the same games, though usually by different scores. Gould has argued that if we ran the evolutionary clock again, we would get a different result (1989, pp. 293-299, 1994). This may well be true, but we have little empirical experience with multiple evolutionary seasons. We do have a lot of experience with multiple sports seasons. And we know that if we could somehow repeat a season (certainly less of a challenge than repeating the earth's entire evolutionary history), the ball would not bounce precisely the same way every time.

The teams in a sports league play the season to establish the one champion. But as every graduate of the Sportscaster's School of Clichés reminds us, on any given day any team might defeat any other team. The best team does not always win. All the result of a single game tells us is who was the best team on that single day.[27] But that is what makes for the excitement, the drama, the entertainment: our knowledge that, in the end, only one team can win, and we do not know which one.

The objective of any public agency, however, is not to provide entertainment. The objective of any system of educational competition is not to generate excitement as schools compete to determine one, sole champion. The objective of selecting winners and losers is not to create drama as everyone waits for the awards envelopes to be opened and the awards to be announced. The agency's objective is to achieve some public purpose. The school system's goal is to educate all children, not just those children in the few, winning schools.

Any system of competition is useful only if it helps achieve this public purpose. As Clotfelter and Ladd state: "The ultimate test of recognition and reward programs is their effects on student outcomes" (1996, p. 47). Any system of competition is useful only if it helps improve *all* students' abilities. Any system of competition is useful only if it motivates people—administrators, teachers, students, parents, and communities—to help *all* students learn. A system of competition that limits, a priori, the number of schools that can win will not do that. Instead, it will demotivate all those who conclude that their school has little or no chance of winning.[28]

Clotfelter and Ladd note: "The morale problems in programs of merit pay for individual teachers are well known" (1996, p. 46). The losers do not like being labeled losers. But merit reward programs for schools create the same morale problems. The principal of one school whose teachers had worked very hard to win one of their state's awards told Clotfelter: "If they hadn't won, there would have been depression" (1995). Central administrators in that state report that local personnel call up and ask, "How come I didn't win?" This question and the anguish behind it requires these central administrators "to act as therapists" (Clotfelter, 1995). How motivational is a system of competition that creates depression and requires its top managers to devote a lot of time to offering amateur therapy?

Creating losers in sports is fine. Every game creates one winner and one loser—and from that comes the drama, the entertainment. We have the picture not only of Bobby Thompson being mobbed by his deliriously happy teammates. We also have the picture of a dejected Ralph Branca sitting alone on the steps of the clubhouse, slumped over so that all we can see is the back of his uniform with its big number 13. That's drama. That's entertainment.

Creating losers in a school system, however, is not fine. Our objective is not to have only one school win. Our objective is to have *every school win*. Our objective is not to have only one student win. Our objective is to have *every student win*. Yet too often we create a measurement and motivation system that automatically labels the vast majority of schools, teachers, and students as losers.

Measures Can Discourage Cooperation

Actually, cooperation is not discouraged by the measures themselves. Rather, cooperation is discouraged by competition for limited, individual rewards that are won by ranking near the top of whatever is being measured.

Imagine that two baseball teams that compete in the World Series are not the two teams that performed the best during the regular season and the league playoffs. Imagine, instead, that these two teams are composed of all-star players, the individuals who performed the best at their particular position—pitcher, catcher, shortstop, and so forth—during the regular season according to some scoring system based on hits, fielding percentage, runs scored,and so on. What would happen during the regular season to teamwork? It would, of course, disappear as the players concentrated on their own, individual scores, not on their team's score.[29] This simple change—from team rewards to individual rewards—would dramatically alter the incentives for teamwork.

If you want to foster teamwork, do not limit the number of people who can win rewards. If you want to foster teamwork, do not pit the individuals on the same team against each other. How much teamwork do you get between two gladiators in the pit?

Clotfelter and Ladd write of "the growing recognition that rewarding individual teachers encourages them to compete with one another rather than to work cooperatively" (1996, p. 27). This switch from cooperation to individual competition occurs, however, only if two conditions are met: First, the number of rewards is limited so that not everyone can win. Second, the rewards are won by individuals, even if they are nominally on the same team. If, however, everyone can win one of the rewards by achieving a predetermined goal—or, more precisely if everyone is on a team that can win one of the rewards by achieving a predetermined team goal, and if there are enough rewards for every team to win—then the members of a team will cooperate. Teams may even cooperate with each other. Yet, this is competition, too.

There are, actually, two types of competition: (a) competition against each other, and (b) competition against a goal. In *competition against each other*, individuals or teams compete for a limited number of prizes, which go only to those who score the best. In *competition against a goal*, every individual or team has an opportunity to win the prize if it scores above a predetermined threshold—that is, if it makes its goal. Athletic competition is of the first type and is the most familiar. But this is not the only kind of competition. More importantly, the second type of competition—competition against a goal—can actually encourage cooperation. If every individual, team, or unit can win one of the rewards by achieving a predetermined goal—and particularly if each individual, team, or unit has a better chance to achieve its own goal if it collaborates with the others—then the competition will actually encourage cooperation.

For educators, there is an obvious parallel. Competition against each other is like norm-referenced testing. Competition against a goal is like criterion-referenced testing.

Simple Measures Can Be Unfair

Measuring the performance of a school looks relatively easy. Collect the scores of the students on some kind of standardized test and aggregate them. Sure, most tests have problems with either validity or reliability. But if we want to learn if children can do arithmetic, we can give them arithmetic problems. We have been doing it for eons.

What if some children come from wealthier homes? What if some children come from homes where the parents buy them books? What if some children come from homes where the parents read at least as much as they watch television? What if some children come from homes where parents read to them? What if some children come from homes where the parents help them study arithmetic? We are looking to measure the teaching performance of the schools, not the wealth of the parents, or the teaching performance of the parents.[30] Just using the students' absolute test scores produces a very predictable result: The schools in the wealthiest communities win; the schools in the poorest communities lose. Surprise?

There are two issues of fairness. The first is whether measurements themselves are fair. Unfair measures will not help managers evaluate or learn. The second issue is whether the measurements are perceived as fair. For any perception that the measures are unfair will limit the managers' ability to use them to motivate performance or to promote accomplishments.

Ladd notes that she and Clotfelter worry about "perceptions of fairness: an accountability system would be deemed unfair if it typically favored schools serving one type of student rather than another" (1996b, p. 13). Specifically, the biggest fairness problems, either real or perceptual, could occur in two situations: (a) if the measurements favored schools serving white students over schools serving minority students, or (b) if the measurements favored schools serving rich students over schools serving poor students. Clotfelter and Ladd argue that "any measure of school effectiveness that is highly correlated with the racial or economic composition of the student body will be difficult to defend" (1996, p. 39).

Measuring Today's Status Tells Us Where We Are, Not How Much Progress We Have Made

During a political campaign, a candidate's pollsters and strategists worry less about what percentage of voters say they will vote for the candidate today and more about how that percentage changed since the last poll. To figure out what to do next, campaign managers need to know how much progress they have just made or not made, how much impact their latest strategy and tactics have had.

The same applies to public managers and to the measurement of school performance. High test scores are certainly good, unless they reflect a slippage from previous scores that were even higher. In this case, the data suggest that the school's principal and teachers need to rethink their current teaching strategy or at least the implementation of that strategy. Low rates of daily attendance are certainly bad, unless they have shown continuous improvement over the past eighteen months. In this case, the data suggest that the current effort to improve daily attendance is working. Public managers need to undertake measurement not only to generate data about existing performance. They need to collect these measures periodically to reveal the nature of any progress.

Complex Measures Do Not Motivate Anything

Dallas, report Clotfelter and Ladd, has sought "to be scrupulously fair to schools" by adjusting a school's score for the demographic, social, and economic characteristics of its students. Unfortunately, the Dallas approach "is incomprehensible to most participants in the process and to most outside observers." As a result, the measure used by the Dallas school system to evaluate its schools cannot motivate improvement. As Clotfelter and Ladd write, "School officials neither understand the process nor have any idea what sorts of gains would have been required for them to achieve a high ranking" (1996, p. 35).[31]

The same problem exists in Kentucky. Elmore, Ablemann, and Fuhrman (1996) report that "many educators and members of the public do not understand the index of school performance that the state uses" (p. 76). They quote one state official as saying: "It's too complicated. That's one of the most serious problems of it, and one of the reasons it's not trusted ... is that people do not understand it" (p. 76). Both Kentucky and Mississippi use "formulas and calculations that are not clearly understood, which makes it difficult for such measurements to motivate useful reform" (p. 77).

Two of the reasons that public managers measure performance are to evaluate and motivate. Yet, there is no reason to believe that the same measure will accomplish both purposes equally well. An evaluative measure has to be fair; if the measure is to reveal how well a public agency is doing, it has to treat all districts equitably. If one unit scores higher than another, it ought to be performing better. Every unit ought to have the same opportunity to score high. The measure ought not to give any kind of unfair advantage. For evaluative purposes, therefore, we create measures with a variety of complex adjustments designed to compensate for a number of potential sources of bias.[32]

To motivate improved performance, however, a measure need not necessarily be perfectly fair, unless, of course, its unfairness convinces people and teams that trying is futile or even dishonorable. A motivating measure, however, does have to be clear. A motivating measure has to send clear signals about what people need to do to improve. A motivating measure ought to motivate everyone. To motivate improved education, a motivating measure ought to be so transparent that everyone—principals, teachers, students, staff, parents, and citizens—knows exactly who needs to do what next to improve the school's performance.

Public managers measure performance for at least five different reasons to help them with at least five different managerial functions. There is no reason to believe that any single measure will be best for all these functions. There may often be a significant trade-off between measures that help managers evaluate fairly and measures that help managers motivate effectively. Complex measures may generate equitable evaluations but have little motivational impact.[33] Simple measures may be constructive motivators but produce biased evaluations.

Measures Encourage the Hoarding of Limited Resources

Even if the measures are very fair or are perceived as fair, even if everyone can win, a performance-driven system still encourages people and teams to accumulate resources. In particular, such a system of competition encourages the accumulation of those resources that are important to success and most difficult to locate and acquire. To the extent that measures are linked to rewards or punishments, organizations will seek the resources they need to do well by these measures. Even if the competition is against a goal, even if every team can win, teams will hoard those key resources that are essential for winning.

For the measurement of performance in elementary and secondary education, these key resources could include good teachers, good students, and good parents. Some district superintendents or school principals might decide that the best strategy for achieving their goal is not to improve the capabilities of their existing teachers, but to recruit the best teachers. Even though every district or every school could win—even though each district or school was competing to achieve its own goal, not to beat the others—the districts and schools would still be competing against each other for resources.

Of course, this competition for resources does generate some helpful learning. It reveals what resources are essential or, at least, are deemed essential. It reveals which of the many versions of these essential resources are best. For example, the competition among schools for teachers can do more than the principals' formal teacher evaluations to reveal to the district superintendent which individual teachers are really the best. The same competition could also reveal what teacher characteristics are necessary to achieve the next goal.

If measures are to motivate improved performance, those managing the performance-driven system will have to make sure that key resources are available to all. They will have to ensure that no individuals or teams can gain an unfair advantage by monopolizing control of some essential but limited resources.

Measures Themselves Cannot Improve Performance; Organizations Also Need Resources

Measures can motivate performance; they cannot produce it. The best-motivated students cannot do their math homework without pencil and paper. The best-motivated teachers cannot get their students to use mathematics to think analytically if they do not know how to undertake this kind of sophisticated teaching or if they, themselves, cannot use mathematics to think analytically.

If measures are to motivate improved performance, those managing the performance-driven system need not only to ensure that all individuals and teams have adequate resources. They also need to ensure that everyone has the knowledge and capability to use these resources. Improving performance often requires significant training devoted to improving organizational and individual capabilities.

A DOZEN HINTS FOR DESIGNING A SYSTEM THAT LINKS MEASUREMENT TO MOTIVATION

Given all these problems, is it possible to use measurement to motivate? Or, is the theoretical link between measurement and motivation too difficult, if not impossible, to activate? The answer, I think, is that it is possible—in practice as well as in theory—to employ measurement to motivate improved performance.[34] Top leadership needs to be careful, however, to create a measurement system that reflects the realities of organizational circumstances and human behavior.

Start With a Few Easy Victories

None of this will be simple. There will inevitably be mistakes along the way. So, do the doable first. Adopt "a strategy of small wins" (Weick, 1984). Get some successes, reward them, learn from them, convince the skeptics that the measurement initiative is valuable and here to stay, convert a few of the real cynics, and then tackle harder problems.

For example, start by measuring the accomplishment of noncontroversial purposes. Do not start by attempting to measure how well the schools are increasing the students' self-esteem or their knowledge of practices for safe sex. Nothing will destroy a measurement program more quickly than attempting to measure the achievement of purposes to which there exists organized, aggressive opposition. Instead, begin with purposes that most people accept as overwhelmingly proper. That is undoubtedly one reason why measurement in mathematics has progressed further than for most other subjects. Most of us agree that two plus two equals four, and that all students ought to know this. Hanushek and his colleagues suggest that at the center of national educational purposes should be "fundamental cognitive skills: high levels of literacy, analytical ability, and scientific knowledge combined with the ability to reason abstractly" (1994, p. 165).[35]

Start with some early victories. Design your initial efforts to be successful. Get a quick and easy win, maybe a number of such wins. Use these victories to convince the skeptics of the value of measurement. Make sure that you set up the motivations for the next stage.

Choose Simple Measures That Will Motivate the Behavior Needed Next

Start by thinking about motivation instead of measurement. After all, when most of us hear "measurement," we implicitly, unconsciously think "evaluation." This creates one image of what kind of measures we need. But, if we are really looking to motivate, we may need very different kinds of measures.

For example, if the purpose of a measurement is to compare and evaluate schools, then a complex measure may be necessary. To capture all of the different things that we want schools to accomplish and to compensate for all of the differences among schools, we need to measure a lot of different variables and incorporate them all into our measure of performance.

If, however, the purpose of a measurement is to motivate behavior, that measure needs to be simple. People need to understand it. More significantly, people need to understand what exactly it measures. They need to be able to figure out what they and their school need to do to make this measure go up for their school. If a measure is to motivate, it should not obscure the behavior that is sought. To motivate, the measure should make it very clear what output we want the school to produce this year, even if it is not immediately obvious exactly how to produce it.

Give Everyone a Chance to Win

We want every school to educate its students. We want every school to win. Consequently, we ought to create a system of competition that gives every school an opportunity to win. Do not arbitrarily fix the number of winners at a tenth, a quarter, or a third of the schools. Whatever the rewards for success are, make sure that there are enough so that every district and every school can win.

Clotfelter and Ladd write that "If rewards are to be given, then the schools must be explicitly ranked either overall or by type of school, such as elementary, middle, or high school" (1996, p. 27). Not necessarily. Awards can be given to all schools that score above a given, preestablished threshold. In that case, all that must be done explicitly is to divide schools into two categories; those that score above the threshold and those that do not. There is no need to rank the schools, no need to publish each school's score. All that needs to be released is two lists: one list of schools that achieved their goal and the other of the schools that did not.

Reward Progress Not Absolute Results

Why do measures used to evaluate tend to become more complicated? Because we are trying to be fair. Because we do not think it is fair for schools in communities with both low family incomes and a low tax base to have to compete with schools in richer communities. So, we create a handicap system—just like they do in golf—subtracting points from the scores of richer schools or adding points to the scores of poorer schools.

Given that each school, to make progress next year, needs to accomplish something different, why not focus the efforts of each school on its next task? Consider, for example, two schools: The Tom Poorman School with average math scores of 400 out of 1000 and the Woody Rich School where this year's average math scores were 600 on the same scale.[36] To ask both schools to improve their scores next year to 650 is hardly fair. That would not be giving both schools a real chance to win. Rather, for the Tom Poorman School, such a threshold goal would be impossible to achieve.

Consequently, why not have a different goal for each school? For example, each school's math goal for next year could be to reduce by 10 percent the difference between this year's score and some maximum achievable standard of, say, 800. Thus, the Poorman School would have to improve its math score from 400 to 440, while the Rich School would have to improve its math score from 600 to 620. Both would face significant, but different, challenges. Both could win.

Having a threshold goal that increases annually eliminates one of the complaints about adjusting scores for demographic differences. The argument is made that such adjustments imply that our standards for achievement, our expectations for performance, are lower for the students from poor communities than for those from rich communities. Such adjustments mean that schools in poor communities can win with a lower score. But, with an increasing threshold goal, we are applying the same ultimate standard to all schools. We are also being realistic; we accept that it will take the Tom Poorman School longer to achieve this goal than the Woody Rich School. We want both schools to make progress. We want both schools to eventually get to the 800 standard. Therefore, we ask both schools to get themselves closer to the maximum achievable score. We ask each school to eliminate 10 percent of its gap.

Set Stretch But Achievable Goals

We want every school to win. But we also want every school to have to work to win. Winning should not be automatic, even for the best schools. We ought to establish "stretch goals," goals that require schools to stretch themselves to win.[37] Behind the use of stretch goals in business is the assumption that small, incremental goals permit most units within a firm to achieve their annual goal simply by working a little harder or a little smarter. Stretch goals are designed to get people thinking differently, to rethink how they perform their basic functions (Tully, 1994). Most schools are not apt to improve their math scores from 400 to 800, or even from 600 to 800, without redesigning their strategy for teaching math.

The first year's targets, of course, might not really be stretch targets. You need some winners in that first year. Remember: Start with a few easy victories. You need to convince people that they can win. You need to convince people from both rich and poor communities that their schools can win. For the first year of this measurement and motivation program, each school's goal might be to close its gap by only five percent. (In the first year, creaming is fine. It helps people learn and builds commitment.) In the second year, the goal might be to close the gap by eight percent. Then, in the third and subsequent years, the goal could be to close the gap by the full 10 percent.

Do Not Rely Upon the Same Measure to Motivate all Schools

Usually, different schools, even within the same district, are performing at different levels. To improve this year, different schools need different performance targets. They will also need to do different things, perhaps fundamentally different things. There is no reason to assume that one single measure will be ideal for motivating these different behaviors. It may not make sense to attempt to motivate the behavior of all schools by using some kind of least-common-denominator measure. For example, some schools may have daily attendance rates that are so poor that efforts to improve instruction in mathematics will hardly affect test scores. For such schools, the first year's goal might have nothing to do with mathematics scores; rather, it might be to increase the daily attendance rate to a maximum achievable standard of, say, 95 percent.

Using daily attendance to motivate the performance of schools has some obvious weaknesses. A school can achieve this goal with a little honest cheating. It can get the students into the school and still teach them absolutely nothing.

If we believe that schools do educate, however, we have to get the students there before the schools can do this job.

Using daily attendance to evaluate the performance of schools also has some obvious weaknesses. Daily attendance is certainly not an outcome measure. It is not even an output measure. It is really only an input measure. It measures how many students we put in the school, not what we accomplish with them once we get them there. Nevertheless, despite its weaknesses as an evaluating measure, daily attendance may be very desirable as a motivating measure. That is, by measuring and rewarding daily attendance, we may be able to motivate precisely the kind of behavior that some schools need to improve performance.

For a school with low attendance, getting more students into the school is a significant challenge. To improve daily attendance figures, the leaders of the school will have to do a lot of things that we want them to do. They will have to make the school more attractive to delinquents. They will have to work with parents and community leaders to drive up attendance, perhaps even making changes to win the cooperation of parents and community institutions in their attendance drive. They will have to be creative in getting students to come to school. They might even try to make mathematics more interesting. Daily attendance is a poor evaluative measure—a poor outcome measure, and a poor output measure. It may, however, be an excellent motivating measure to drive an entire variety of desirable behaviors.

Of course, merely improving attendance is hardly an accomplishment. The students may be at their desks, but they may not be learning enough to even be promoted to the next grade. Consequently, after a school has eliminated the attendance gap, after a school has reached the maximum achievable attendance standard, it still may not be ready to tackle the mathematics goal. It might be better to motivate this school to focus on increasing promotion or graduation rates. The concept would remain the same. Establish a maximum achievable standard for promotion. Give each school the new, annual goal of closing the gap between this standard goal and their own promotion rate by 10 percent. Reward each school that achieves this annual goal until it has reached the standard, as long as it also continues to achieve the maximum achievable standard for daily attendance. Then, promote the school to a new performance measure, maybe mathematics, and use it to motivate behavior.

All this need not be very complicated. The strategy involves not a single goal but a hierarchy of goals. The first goal in this hierarchy might be to get daily attendance above 95 percent if a school did not achieve this level of performance last year. If attendance was only 75 percent last year, then daily attendance becomes this year's performance measure and, say, 80 percent attendance becomes this year's goal. Once a school achieves the 95-percent daily-attendance goal, that becomes part of its continuing annual performance goal along with a second performance measure, such as increasing the grade promotion rates or decreasing the dropout rate. To be successful in any given year, a school must improve its performance on its current measure, closing the gap between last year's measure and the maximum achievable standard by 10 percent. It must also maintain its performance at the maximum achievable standard for all the lower goals in the hierarchy.

Some schools need to solve very basic problems; they need to drive up attendance rates, grade promotion rates, or graduation rates. Other schools may have already solved such problems and now need to get their students to analyze complicated problems and to explain their conclusions and reasoning in a logical essay. We may need to motivate different schools to do different things. If so, we need to use different motivation measures.

Such a hierarchy of goals motivates each school to improve its performance. Such a hierarchy of goals will give each school an opportunity to accomplish something significant each year. Such a hierarchy of goals permits every school to win every year. Such a hierarchy of goals does not discriminate against poor schools. Such a hierarchy of goals gets each school to concentrate its energies on what it needs to do next.

The schools that we want to motivate the most are the poorest performers. The schools that most need to develop a sense of accomplishment, a sense that they can improve are, again, the poorest performers. The schools whose performance we most need to continually ratchet up are the poorest performers. Any measurement and motivation system that fails to motivate the poorest performing schools, that fails to let these schools develop a sense of accomplishment and the confidence that more improvement is possible, and that fails to continually improve their performance is not designed to really improve a state or district's overall performance.

Drucker has suggested a number of "deadly sins in public administration." One such "strategy guaranteed to produce non-performance is to try to do several things at once" (1980, p. 103). Instead, Drucker argues that improved performance depends on concentrating on a priority. Creating a single performance goal to motivate a school's behavior establishes this priority. But, if different schools have already achieved different levels of performance, they need different priorities, different goals. By creating a hierarchy of goals for schools, a district can be sure that each school is concentrating on a priority that is most appropriate for its level of past performance and its need for future improvement.

To Motivate Team Behavior, Reward Team Behavior

Twenty years ago, Kerr wrote about "the folly of rewarding A, while hoping for B." Reflecting on the kind of management systems employed by both public and private organizations, Kerr observed: "Managers who complain that their workers are not motivated might do well to consider the possibility that they have installed reward systems which are paying off for behaviors other than those they are seeking" (1975, p. 781). The general lesson is clear. The specific lesson for people who seek to encourage cooperation among individuals ought also to be clear: If you want people to work cooperatively as a team, reward them as a team, not as individuals. It is pure folly to reward individuals while hoping for teamwork.

Ladd argues that "the school is the most appropriate unit of accountability" (1996b, p. 11). Certainly, we want to encourage cooperation among teachers and others associated with a particular school. So, give rewards to schools rather than to individual teachers. Or, if we give out individual rewards, give them to all the teachers in each school that makes its collective performance goal. Dallas, for example, gives bonuses to all teachers and staff in a winning school (Clotfelter & Ladd, 1996, p. 35).

We also want to encourage cooperation among schools, not just within schools. So, why not give rewards to districts as well as to schools. Measures need not be collected only for teachers, only for schools, or only for districts. We can collect measures for all levels in a state school system. Similarly, rewards need not be limited to some levels. We can offer rewards for the achievement of goals at all levels. For example, a district might win a particular award when all of its schools reach a particular maximum achievable standard in daily attendance or when it closes its promotion rate gap by 10 percent.

Do Not Rely Upon Monetary Rewards

One reason for fixing the number of winners is that funds available for rewards are limited. If too many schools win, the rewards will have to be spread around and will be too small to motivate anyone. Given the turnover in legislative membership and the instability in legislative preferences, it is often difficult to sustain a level of monetary rewards over a period of many legislative sessions. Legislators are attracted to the common-sense theory of pay for performance; they are less attracted to the practical reality of funding it.[38]

Even if monetary rewards are, or can be, made to be effective motivators, even if they are always available to public managers, why should we limit our motivational strategy to one motivator? Why not employ motivators that build upon an individual's need for love, belongingness, esteem, and even self-actualization—upon the entire hierarchy of people's needs? Why employ only one motivator when other effective ones are also available? Why rely solely on motivators that are the most subject to legislative tampering and legislative revocation? Why not employ rewards that are cheap, yet effective? Public managers who seek to motivate improved performance and school officials who wish to motivate improved school performance ought to invent a variety of ways to use psychological as well as material motivators.

Reward the Winners By Asking Them to Help the Losers

Any system of measurement and motivation needs feedback. For measurement to drive improved performance, it also has to drive learning. In particular, it needs a mechanism to help the poor performers learn from the best. It makes sense, therefore, to give particular winners the task of each helping a particular loser to learn. Give the very best schools, principals, and teachers the assignment of assisting the schools, principals, or teachers who are having the most difficulty achieving their goals. This creates a feedback system that encourages learning.

At the same time, it motivates. You are not giving these winners more money; instead, you are giving them more work. In the process, you are also converting them into role models, leaders, and heroes. This is the reward.

These assignments need not go to those schools with the very highest scores. The scores, after all, do not even have to be known. Rather, this is a way to insert some intelligent, albeit subjective, judgment into the reward process. This is a way to reward the very best performers without producing resentment.[39] It is also a way to help improve the performance of those who need it most.

A system of measurement and motivation need not rely on just the formal announcement of this year's successful schools to recognize and motivate performance. There are many other, more subtle ways to reward really outstanding performance. One is to use the task of helping the losers to recognize the really top achievers, to broaden their circle of belongingness, to enhance their esteem, and maybe even to provide new opportunities for self-actualization.

Pay Attention to the Details

As Hanushek has emphasized, "The details matter a lot" (1996, p. 128). If we are to use measures to motivate, we need to think carefully about the specific measure we are using and how people will respond. How can they engage in honest cheating? How might they engage in dishonest cheating? Hanushek and his colleagues note the importance of the details in shaping incentives:

> The effectiveness of an incentive system depends directly on how individuals act when faced with a given pattern of rewards. So rewards and punishments must be sufficient to get people to change their behavior, and they must be precise in their effect. Impact and effectiveness flow from the details (1994, p. 90).

You not only have to get the macro motivational theory right, you also have to get the micro motivational details right.

For example, what number should we use to measure the aggregate performance of each school's students on a standardized mathematics test? The mean? The median? Does such a detail make a difference? Absolutely.

Automatically, yet only implicitly, we usually assume that the score to best capture each school's performance is the average.[40] We do not even stop to think of the motivational implications of using this most common arithmetical formula. We automatically assume that a school's score is the average of its students' scores. And, for some evaluative or comparative purposes, this might be correct or at least acceptable. Yet, this is not the only choice of a measure of the school's performance; we could use the median rather than the mean. Moreover, slightly different measures can motivate significantly different behavior.

South Carolina, for example, has chosen to use the median gain of a school's students as a measure of that school's performance. This, note Clotfelter and Ladd (1996), has advantages and disadvantages:

> The use of the median keeps teachers from focusing attention on students they perceive as fast learners who could make the school look good. However, it also gives them little incentive to give attention to students they perceive as slower learners (p. 31).

If the mean student gain is used to measure a school's performance, the same gain of any student's score contributes the same amount to an increase in the school's score; thus, schools and teachers are encouraged to concentrate on the students with whom, they believe, they can produce the biggest gains. If the median student gain is used to measure a school's performance, the only gains that matter are those of the students at and around the fiftieth percentile. Schools and teachers, therefore, are encouraged to concentrate on the students whose scores, they believe, will be near the median. For most schools most of the time, the difference between the mean and median scores will be very small. Yet, the choice between them as a measure of school performance has large motivational implications.

What kind of behavior are we trying to motivate? Do we want schools to ratchet everyone up a little? Or, are we really concerned that schools are failing those students who are doing most poorly? If so, why not measure a school's performance, not by using its mean or median score on some standardized test, but by using its tenth percentile score? What's magic about the average except that we use the word "average" in our daily lives and learned to calculate it in the eighth grade? What's magic about using the mean, or median, or some other more complicated measure of central tendency? Is the school's average grade really the best measure of its performance? Not if we really want to motivate schools to focus resources, attention, and teaching on those students who need the most help.

Measuring a school's performance with its tenth-percentile score can motivate very different behavior than using its average score. With the average score, the increase in every student's score counts the same; consequently, the school's math department will be encouraged to focus its attention on those students most capable of improving their scores. With the tenth-percentile score, however, improvements by the best students count for nothing. Rather, the improvements that really count are those of the students scoring around the tenth percentile, the students in the lowest quintile.[41] These are the students on whom the math department needs to focus its attention.

When a state or school system chooses from among the array of available standardized tests the one that it will use to measure student and school performance, it is not finished making the decisions about the measures it will use. For each school will generate a distribution of student results. How should this distribution of individual students' performance measures be aggregated to create a measure of the school's performance? How is the distribution of student scores to be employed to create a school score? There are a number of alternatives. Calculating the average is only the obvious answer, not necessarily the best. How the measure of school performance is created will have some important motivational implications. If, for example, the measure is chosen carefully and communicated intelligently, it might even motivate the students who do best in math to tutor those having the most difficulty. Such a measure of mathematical performance might also have important consequences for teaching citizenship.

Experiment With the Details

Will measuring, reporting, and rewarding a school's tenth-percentile score on a standardized math test create different incentives than measuring, reporting, and rewarding the school's average score? Obviously. Will it create better incentives? Perhaps. That, however, is not guaranteed. How people will respond to different measures cannot be predicted perfectly. We will not really know until we try it.

"God is in the details," said the architect Ludwig Mies van der Rohe. But in which details? "Discovering those details should be the focus of experimentation," argue Hanushek and his colleagues. "Much more experimentation will be required to find performance incentives that perform" (1994, p. 90). We will never know which precise performance measures create which precise incentives until we try them.

Unfortunately, such experimentation, note Hanushek and his colleagues, "presents political difficulties for legislators, for in pursuing it they must accept that nobody yet knows the details of programs that will improve schools." And, of course, no one campaigns for the legislature by explaining to the public that he or she does not have the answer. Yet, continue Hanushek and his colleagues, experimenting "means that instead of selling voters a solution to the problems of the schools, legislators must convince voters to embark on the search for a solution, a search that will require considerable effort, and some risk, on the part of voters themselves" (1994, p. 116).

What is needed is a strategy that might be best described as "management by groping along" (Behn, 1988; Behn, 1991, chap. 7). We know what we want to accomplish: to create a system of measurement and motivation that drives all schools to improve the performance of all their students. But we do not know how to get there. So, we experiment; we grope along. We try some things, see what works and what does not, make some changes, and try again.

This is the same strategy that individual schools will have to follow. They, too, know, or should know from the system of measurement and motivation, what they want to accomplish: to improve the performance of all their students. But they, too, do not know how to get there. They must grope along.

Monitor Many More Measures Than You Use to Motivate

To motivate behavior, it is usually best to use just one measure of performance. Employing a single measure eliminates confusion. It eliminates Drucker's sin of trying to do several things at once. It establishes a clear priority. It focuses people on producing one, specific result.

But that single measure of performance can also distort behavior. It can encourage both honest and dishonest cheating. It may produce unintended and unpredicted consequences. The managers of any system of measurement and motivation need to know about these results, too. They need to be able to evaluate and to learn. To do this, they need to analyze a lot more data.

People need to monitor a variety of other performance measures. These should include measures that can help with the tasks of evaluation and learning. Are some schools that meet their math targets, for example, also doing better in reading comprehension? These measures might include some that are closer to being real output or outcome measures. They ought to include measures of performance that we anticipate or hope will result from people's efforts to improve against the single, motivating measure. Will improving daily attendance, for example, encourage schools to make math more interesting and also improve students' math skills? They certainly ought to include measures that could identify likely methods of cheating, both honest and dishonest, and that could catch the cheaters, particularly the dishonest ones.

Measuring performance has lots of purposes. To accomplish the motivation purpose, to drive the behavior of organizations and individuals, one specific motivation measure is most helpful. But to accomplish the other purposes, to evaluate, promote, learn and celebrate, it helps to monitor a lot of other performance measures.

Try to Create Long-term Support

Performance measurement might be the hottest topic in government *today*. But what about *tomorrow*? Is this just another management fad to be quickly discarded when another management guru publishes another best-selling book?[42] Or, are the fundamental principles behind performance measurement sufficiently sound to outlast political and intellectual trendiness, to say nothing of the personal preferences of legislative and executive branch leaders?

Why should people take any system of performance measurement and motivation seriously? If the program is to be short-lived, why make the effort? Why should they exert all the energy necessary to radically alter their behavior and their organization's behavior simply for some short-run praise? Based on their past experience, people will worry that this program also will soon disappear.

Actually, what they may fear is that half of the program will soon disappear. That is, they may worry that the rewards will soon disappear. Only the measurements and the punishments will continue. That would be the worst of both worlds.[43]

Consequently, those who seek to create a system of measurement and motivation need to create a structure of political support necessary to sustain the system over time. After all, the impact of any such system, the impact of any system designed to affect the behavior of professionals and institutions, will not be felt for years. "The three years since the beginning of the program" in Dallas, write Clotfelter and Ladd, "is probably too short a time for the program to induce the types of change in culture, personnel, and teacher training necessary to promote and sustain long-term gains in test scores" (1996, p. 47).

A continuous stream of small victories can help to create political support. If district superintendents, school principals, and teachers are winning and understand how they can continue winning, they may support rather than oppose the system of measurement and motivation. If this system generates a continuous stream of small but accumulating victories, it will serve to promote not only the winning school but also the measurement and motivation system itself.

THE BIG QUESTION ABOUT MICROMANAGEMENT

My third big question about public management concerns micromanagement:

> How can public managers break the micromanagement cycle—an excess of procedural rules, which prevents public agencies from producing results, which leads to more procedural rules, which leads to ...? (Behn, 1995, p. 315).

Micromanagement plagues the delivery of every public service, including education.

Educators are not only told what to produce in terms of test score results, they are told how to produce these results through uniform rules that regulate everything from the credentials of teachers to the content and pace of the curriculum. The rules about how schools should produce results are often more specific and more watched than the actual results these schools are supposed to produce by following all the rules.

In the concluding chapter of *In Search of Excellence*, Peters and Waterman (1982) argue that one of the traits of excellent businesses is their "simultaneous loose-tight properties" (p. 318). High performing firms are tight about what they want to accomplish but loose about how they accomplish it. They drive their divisions and other subunits to achieve specific goals, but they give these units considerable flexibility in achieving these goals. They are tight about ends but loose about means.

Government, in contrast, manages with simultaneous tight-loose properties. Too often, government is loose about ends but tight about means. Too often, government cannot agree on what goals it wants its agencies to achieve, but it can dictate to the finest detail how they should do it. Government agencies are charged with accomplishing major, vague, and often conflicting purposes, yet they are given little flexibility. It is micromanagement at its worst.

Ladd writes that "the new educational accountability is shifting from input and process to performance as the basis of accountability" (1996b, p. 6). Perhaps the use of measurement to motivate improved performance reflects a shift toward a more simultaneous loose-tight strategy for managing government. Still, any increased flexibility needed to really improve performance will have to be earned.

The political culture that tightly regulates managerial flexibility, that creates new rules to prohibit the repetition of past abuses, is endemic to American government. Changing such tight-loose habits will not be easy. There still are political points to be earned by championing a rule to prevent, once and for all, some large mistake or small indiscretion. American politicians and American citizens will not easily give up their love for tight rules, their desire to ensure that public managers and government employees always do precisely the right thing.

The strategy of small wins may help to earn the flexibility necessary to achieve bigger victories. Public managers measure performance not only to evaluate and motivate, but also to promote. A series of annual successes, as proven through a series of annual measures, can promote the value that an agency is producing. Such annual successes can promote the need for more flexibility to achieve even bigger successes. Such annual successes can promote the organization's competence, demonstrating that it is worthy of trust and of a grant of greater flexibility.

To improve performance, to produce real educational results, a school system needs to simultaneously be both loose and tight. It needs to be tight about its purposes, about the goals each school needs to pursue next, and about measuring the achievement of these goals. Yet, it needs to be simultaneously loose about how a district, school, or teacher is to accomplish these goals. By employing measurement to demonstrate the achievement of tight, educational purposes, a school system may be able to earn a little more looseness.

MEASURING, MOTIVATING, AND PRODUCING PERFORMANCE

It is one thing to motivate performance—to stimulate people to attempt to improve the effectiveness of schools—but it is quite another to actually produce that performance. Measurement is not enough.[44] If that were true, the mere publication of state report cards would have caused student test scores to jump dramatically. Motivation is not enough. As Ladd observes, "Administrative accountability systems and incentives systems are not a panacea" (1996b, p. 14). Even flexibility is not enough.

Schools need resources. That means more money. Yet, money is not enough either. Schools also need institutional and human capabilities. They need to not only know what they want to do and to have the financial resources, flexibility, and knowledge to do what they want to do, they also need to have the physical, mental, psychological, social, interpersonal, leadership, and organizational capabilities to accomplish their purposes. Measurement and motivation can provide the framework in which these capabilities can prove effective. But these capabilities are necessary, too.

NOTES

1. As mayor of New York City, Ed Koch did precisely this. He was very concerned about whether the city's agencies paid their bills on time. At meetings of his department heads, he would distribute a listing of city departments ranked by what percentage of the agency's bills had been paid within 30 days.

2. *The 1993 report card: The state of school systems in North Carolina* issued by the State Board of Education was prepared in the Department of Public Instruction by its Office of Accountability Services. I wonder how many of the state's 120 school districts thought that this office was providing them with an accountability "service"?

3. What does this say about the value of regulations when freedom from regulation is used as a reward? Are regulations just another form of punishment, like regulating the television-watching hours or the Saturday night curfew time of ill-performing teenagers? If removing regulations is a reward, why do we punish schools in the first place by giving them so many regulations?

4. Hanushek and his colleagues (1996) argue that schools do not use data to learn: "Today America's schools only teach. They themselves have little motivation to learn. Working and workable programs are difficult to find because schools all too often fail to monitor systematically the success or failure of programs now in place" (p. 55). In fact, they devote an entire chapter to "Turning Schools Themselves into Learning Institutions," emphasizing how schools can use measures to learn (pp. 125-149).

5. Nevertheless, Pillsbury includes "celebrate" as one of her five reasons for measuring performance because it is so often overlooked. Pillsbury is big on celebrating accomplishments (Behn, 1991, pp. 104-117).

6. In an examination of measurement problems in education, Wainer implicitly focuses on an entirely different purpose: predicting which students will do well at the next level or on the job. Wainer is interested in "measurements that lead to a decision and a consequence" (1993, p. 2) though these are not policy decisions or policy consequences but admission and hiring

decisions that have consequences for the organizations making these decisions and for the individuals applying for admission or employment. Although this predictive purpose is certainly a valid one for the measurement of a student's proficiency in a subject—particularly for someone like Wainer who is employed at the Educational Testing Service—they are much less important for managerial purposes.

7. Hanushek and his colleagues apply Chase's principle to individual students:

> Individual assessments also create incentives that motivate students to study. A key factor in the strength of those incentives, however, is the degree to which assessment information is made public and used to judge the student. Promotion through the grades, graduation, and the like constitute an important part of the public's general knowledge about individual students, providing strong incentives for them to perform well in those terms. Scores on standardized tests and course grades are sometimes used to make judgments about students, for example in the college application process, but frequently they are not made generally public, thus lessening the incentives for students to perform well (1994, p. 128).

Hanushek and his colleagues appear to be even more direct than Chase; it almost seems as if they would post the ranked results of every test on every school bulletin board and maybe in the local newspaper.

8. This suggests that a measure that is never published comparatively has no motivational consequence. If you received a grade of Δ, but have no idea what grades others received or what a grade of Δ means, the grade would have no motivational impact. To activate the measurement-motivation link, there has to be—at minimum—a psychic reward that comes from knowing that you are one of the winners in the crowd.

The impact of psychic rewards should not be underestimated. Just think of all those bumper stickers that you see announcing: "My child is an Honor Student at Jackie Robinson High School." Except for principals who assign their teachers to run around the parking lot before and after school pasting those stickers on the right parents' cars, those bumper stickers were put there by the winner's parents—by proud parents. Whether they had anything to do with their child's success or not (and we all know that they had at least something to do with it), they are proud of this success. They like being recognized. They like having their child recognized. They like being associated with success.

The announcement or publication of a list of winners tells you three things: (a) It tells you how well you are doing, (b) it tells you how well everyone else is doing, and (c) it tells you that everyone else knows how well you are doing. When the Massachusetts commissioner of public welfare published the error rates of each office "things changed" not only because "everyone knew if his or her office had a high rate" (Osborne and Gaebler, 1992, p. 146). Things changed because every office director knew the rate that every other office had. Moreover, things changed because every office director knew that all the other directors knew exactly what his or her office's rate was. These three pieces of information, particularly the last one, provide the psychic reward that links measurement to motivation.

9. When Hanushek and his fellow economists first discuss performance incentives, they specifically note that money is not the only possible reward:

> The rewards to teachers may be explicit monetary rewards, or they may be a wide variety of intrinsic rewards, such as special recognition, more latitude in classroom and activity assignment, or expanded travel and training opportunities (1994, p. 87).

Yet, later, when they get into more detail (pp. 118-121), they write as if money is the primary reward. They observe, for example, "that no existing system of school finance in America explicitly rewards performance. No state gives additional funding to a district simply because student achievement has improved" (1994, p. 120). They never say that money or salary or funding is the only reward; yet, that appears to be implicit in much of what they write. I think most of us, when we think about providing rewards for performance, think implicitly and exclusively about extrinsic rewards, specifically monetary rewards.

10. Herzberg (1968, p. 59) offers seven different motivational strategies:

1. Removing some controls while retaining accountability.
2. Increasing the accountability of individuals for their own work.
3. Giving a person a complete natural unit of work (module, division, area, classroom).
4. Granting additional authority to an employee in his activity; job freedom.
5. Making periodic reports directly available to the worker rather than to the supervisor.
6. Introducing new and more difficult tasks not previously handled.
7. Assigning individuals specific or specialized tasks, enabling them to become experts.

Given the nature of teaching, employing these strategies might require only a little creativity.

11. Maslow calls this "a hierarchy of relative prepotency" (1943, p. 375).

12. Money can buy love, affection, belongingness, self-esteem and the esteem of others, but only to a limited degree. Obviously, money can buy membership in many (but not all) groups. In American culture, of course, possessing money is a mark of an individual's economic and social worth and, thus, a source of esteem from others and from one's self.

Money can directly buy esteem. Why else do people make those large contributions to universities? An anonymous contribution may well be motivated only by self-esteem or by the self-actualization of doing something creative with one's resources. In contrast, a very large contribution buys not only a new campus building, but also the esteem that comes from having your name carved into the front of the building for others to see, as well as from having all of the speeches delivered in your honor at the building's dedication.

Still, not all money can buy love or esteem. Universities have on occasion (honest, I am not making this up) refused large donations because of the way in which the money was obtained.

13. Note that a requirement that teachers teach a predetermined curriculum in a predetermined way undercuts their ability to pursue self-actualization.

14. Maslow emphasizes that to move up his hierarchy from one need to the next does not require complete satisfaction of the first need. Most people, he writes, "are partially satisfied in all their basic needs and partially unsatisfied in all their basic needs at the same time" (1943, p. 388). But the lower needs are more satisfied than the higher needs, and, as an individual's lower needs become more satisfied, he or she turns more attention to the fulfillment of higher needs. Thus, the easier it is to satisfy an individual's physiological and safety needs, the quicker that individual will focus on the needs of love and esteem.

15. The public manager's advantageous selection is the opposite of the economist's adverse selection.

16. This is the theme of Kohn's (1993) *Punished by rewards: The trouble with gold stars, incentive plans, A's, praise, and other bribes*. By rewarding excellent performance with bonuses that Kohn calls bribes, we undercut an individual's sense of personal accomplishment that derives from doing something solely for the value produced by doing it. By giving an individual a reward that was promised before the desired act was undertaken, we deny the recipient the right to say that he or she did whatever is being rewarded solely because it was the right thing to do.

17. "Most behavior is multi-motivated," emphasizes Maslow (1943, p. 390). "Typically, an act has more than one motivation" (p. 370). Thus, it might make sense to include money as part of a multi-motivating strategy. That requires that the amount of money be large enough to purchase some significant increase in physiological or safety needs. If, however, the motivational

characteristic of the money is to dramatize our love and esteem for the recipient, why not simply eliminate the money and concentrate on other symbols that can make the love and esteem more real and more widely shared?

18. The GASB suggests measuring the "percentage of parents or graduating students who rate the instruction received as preparing the student to be a responsible member of society" (Hatry, Alexander, & Fountain, 1989, p. 35). At best, this is an output measure. It can only suggest whether the graduating students will, decades hence, be citizens who take their social responsibilities seriously.

19. The idea behind "performance assessments" is that they are much better than standardized tests precisely because they do a much better job measuring success in achieving our true educational purposes and also do a better job motivating teachers to focus on accomplishing these purposes. See, for example, Meisels, Dorfman, and Steel (1995). But how much is better? Is there just less motivational distortion, or is there significantly less motivational distortion from performance assessments? Do the reliability problems associated with performance assessments frustrate teachers more than motivate them? Some are skeptical. Bond argues that "it is unlikely that performance-based assessments will live up to the rosy prognostications of its advocates" (1995, p. 24). And Shepard and her colleagues experimented with instruction motivated by performance assessments and found that "the performance assessments—even with the diligent effort of most project teachers and the commitment of four university researchers—did not automatically improve student learning" (Shepard et al., 1996, p. 17).

20. Ladd argues that "a balance must be found so financial awards are large enough to change behavior, but not so large that they induce outright cheating" (1996b, p. 14). Clotfelter and Ladd make the same point (1996, p. 57). Yet, it is not obvious that people—educators or others—will first change their behavior in the desired ways and only later, as the size of the award is increased, consider cheating. If people conclude that cheating is easier than changing behavior to really accomplish the purpose behind the measure and the award, and particularly if they do not agree with this purpose, they may well select cheating as their first response. The propensity to cheat is not only a function of the size of the financial award, it also depends upon two other factors: (a) the relative effort required to cheat compared to the effort required to actually pursue the purpose behind the measure, and (b) the acceptance or rejection of this purpose as socially desirable and important.

21. When society discovers someone benefiting from a form of honest graft, it converts the behavior into dishonest graft by creating a law, regulation, or rule that forbids it.

22. Rather than using the results of a single test, a state or school district could administer several tests. This could, of course, make it more difficult to teach to a single test. It would, however, encourage teachers to teach test-taking skills. In today's society, this may not be all bad. At least until a person is in graduate school, test-taking skills are important, and test taking strategies and the principles behind them may have other uses in modern society.

23. The number of rewards may be limited by the amount of money available for rewards. Dividing the amount available by the appropriate size for a reward produces the number of rewards to be given. Of course, the amount available could simply be divided by the total number of units in the competition; but because the amount available is not particularly large, the size of the reward would then have to be kept very small. These amounts might be so small as to make the concept of a reward seem silly.

24. How many schools are winners in Dallas each year depends upon the number of people working in the top schools (Clotfelter & Ladd, 1996, p. 35). Essentially, Dallas fixes not the number of schools that will win but the number of school employees who will win.

25. The Dallas program awards those schools that make the biggest gains from one year to the next, making it difficult to be a repeat winner. Dallas employs a sophisticated formula to adjust for the demographic differences among schools. In the first three years of the Dallas program, only four percent of the schools won all three times. Despite Dallas's careful design, more than half of the schools never won (Clotfelter & Ladd, 1996, p. 42). Is this motivating?

26. If only half of the schools in Dallas ever win, maybe the other half will completely drop out of any real participation in the program. If this is the motivational consequence of limiting rewards to only the top-performing schools, it may make it difficult to use the measurement system to promote the schools' overall performance, for it will be more difficult to demonstrate that the program has produced a system wide or state wide impact. Suppose, for example, that to be statistically significant the average improvement required of the students is X. Then, if the students in half of the schools do not improve at all, to demonstrate a statistically significant improvement for the entire system, the students in the remaining, competitive schools will have to improve by an average of 2X. If the students in two-thirds of the schools do not improve, the average improvement required of the students in the remaining, competitive schools would be 3X.

27. On September 14, 1996, the United States defeated Canada 5-2 to win the first World Cup in hockey. With less than four minutes to go in the third period, the United States trailed 2-1, only to score four more goals and take the series two games to one. The result was hardly expected. After all, Canada invented hockey; it is their national game. Dupont summarized the thinking of many observers:

> The victory was dramatic, sensational, but hardly convincing. In truth, the wrong team won Game 3, the series-clincher. Canada was by far the better team, in style of play and execution, for a vast majority of the game....
>
> But the wrong team winning is as much a part of hockey as the net, the puck and everything else.
>
> When it comes to hockey, expect the unexpected, surrender to serendipity. The game's unpredictability is part of its charm (Dupont, 1996).

But it did not take the first 56 minutes of that final game to convince most people that Canada was "by far the better team." Most people thought that from the beginning. Then why play the game? Precisely because using direct competition to select a single winner creates both the drama and the charm.

28. Clotfelter and Ladd note that, in Dallas, "The large proportion of nonwinners means that many teachers may come to believe that they can never win" (1996, p. 46).

29. Of course, to the extent that sports teams reward their players (through bonuses and salary increases) for individual performances, they motivate their players to concentrate on individual statistics rather than on the team's won-lost record.

30. If the school decides that parental involvement is an effective educational strategy, that getting parents to read to their children helps children learn more, and that getting parents to help their children study arithmetic helps children learn more, then we would want to include in our measure the impact of teaching done by parents. We would not do this because we want to evaluate parents but because we want to *motivate* schools to tap this important and effective resource and because we want to *learn* what kind of parental teaching is most effective.

31. Complex measures do not help to promote either. If the public does not understand the meaning of the measures being used, then it will also not understand the significance of a one-year 50-point increase in that measure. Is a 50-point gain big or small? Who knows?

32. This is also why we worry so much about the reliability of tests. For we implicitly assume that the primary, if not sole, purpose of the measure produced by the test is to evaluate. If a test is not reliable, we cannot use it as an evaluative measure.

33. Complicated measures may also be fairer because they are more difficult to manipulate. Write Clotfelter and Ladd: "A sophisticated method of adjusting for student background that counts gains rather than average scores is not as easy to manipulate" (1996, p. 45).

34. I have not seen it work in elementary or secondary education, though I have seen it work in a variety of other public-sector settings.

35. Of course, in American society today, there is some controversy about some of the teachings of science.

36. Between 1880 and 1888, Thomas Iverson ("Tom") Poorman played in the outfield for a number of major-league baseball teams including Buffalo and Boston in the National League and Toledo and Philadelphia in the American Association; in 1887, he led the American Association with 19 triples. Woodrow Earl ("Woody") Rich was a pitcher, appearing in 26 games (1939-1941) for the Boston Red Sox, and in seven games in 1944 for the Boston Braves. In 1941 he appeared in two games, giving up eight hits and seven runs for an ERA of 17.18 (Wolff, 1990, pp. 1344, 2154).

37. The assumption behind this annual goal—closing the gap to the maximum achievable score of 800 by 10 percent—is that as schools get closer to the 800 each point is harder to achieve. If this is not true, the annual goal for each school could be to improve by 30 points annually until the maximum achievable score of 800 is reached. This is essentially the Kentucky approach: Every two years, each school is supposed to close the gap between its original baseline score and a state standard by one-tenth of that original gap. After 20 years, all schools should be operating at the state standard (Elmore, Ablemann, & Furhman, 1996, pp. 70-71).

38. I am indebted to Richard Elmore for this phrase: "common-sense theory." It is a nice way to capture the simple, apparently obvious ideas that we all carry around with us in our heads. Such ideas, which seem so apparently obvious that they need no testing, seem to be completely immune to conflicting empirical evidence. Efforts to motivate organizational and human behavior in the public sector are based on many such common-sense theories.

39. If a few principals, who think they are really hot-shot performers, complain that they have not been assigned to help a needy school, give them an assignment, too. Reward their initiative.

40. Hanushek and colleagues report: "Many states now require periodic examinations using standardized tests. Data on average performance on those tests, broken down by district and often individual schools, are frequently made available to the public" (1994, pp. 132-133). Average scores might be adequate or even good for promotional purposes, but are they the best measures to motivate? What kind of behavior does average performance on a test motivate?

41. Using the tenth-percentile score will not motivate the school to spend any time on the students who are performing in the bottom two or three percent of the population. Every school must have a bottom tenth percentile. It makes no difference whether the fifth-percentile student is almost as good as the tenth-percentile student or gets a zero; if the school's score is the score of the tenth percentile student, the score of the fifth-percentile student, as well as the score of the ninety-fifth percentile student, is irrelevant.

Using the tenth percentile score creates another, perverse incentive: It rewards schools that discourage students below the tenth percentile from taking the test or that encourage these students to drop out of school completely. When the results of these lowest-scoring students are no longer included in the formula used to determine a school's score, this score goes up. But, almost any reasonable method of using individual student scores to create an aggregate school score encourages this form of dishonest cheating.

Using the median score rewards schools that discourage students below the median from taking the test. But, if a school's score is the score of the median student, and if the students' scores are distributed roughly normally, then the scores of those students just above the median will be almost identical to the median score. The scores of those students just above the tenth percentile will be further above it. Using any percentile as the school's score—whether it is the tenth or fiftieth percentile—motivates schools to get rid of their poorest scoring students. The further away from the densest part of the distribution is that percentile used for scoring, the greater will be the impact of getting rid of the poorest scoring students and the greater the incentive to do so.

42. Performance measurement is one of the management fads examined by Walters (1996).

43. This is another reason not to rely upon monetary rewards. Such rewards are clearly dependent upon legislative support. Even if the legislature's leadership believes the monetary

rewards are helpful and important, such rewards need to compete with a lot of other helpful and important uses of limited taxpayer funds. Legislators and legislatures have multiple priorities. It is impossible to guarantee that the level of monetary rewards sufficient to continuously motivate continuous improvement will be continuously available.

44. Hanushek and his colleagues observe that, although measurement, by itself, may motivate people in the right direction, it is not enough to improve performance:

> Some argue that improved student testing and assessment can lead reform by itself. With accurate assessments of student performance from tests, runs the argument, schools will automatically focus on improving test scores; teachers will adjust to improve student achievement, management decisions will become obvious, and reform will occur almost spontaneously. We know of no evidence to suggest this is a realistic expectation (1994, p. 126).

REFERENCES

Allison, G. (1980). Public and private management: Are they fundamentally alike in all unimportant respects? *Setting public management research agendas: Integrating the sponsor, producer, and user*. Washington, DC: U.S. Office of Personnel Management.

Behn, R. D. (1988). Management by groping along. *Journal of Policy Analysis and Management, 7* (4), 643-663.

Behn, R. D. (1991). *Leadership counts: Lessons for public managers*. Cambridge, MA: Harvard University Press.

Behn, R. D. (1994). *Bottom-line government*. Unpublished essay.

Behn, R. D. (1995). The big questions of public management. *Public Administration Review, 55* (4), 313-324.

Blodgett, T., & Newfarmer, G. (1996). Performance measurement: (Arguably) the hottest topic in government today. *Public Management, 78* (1), 6.

Bond, L. (1995). Unintended consequences of performance assessment: Issues of bias and fairness. *Educational Measurement: Issues and Practice, 14* (4), 21-24.

Chase, G. (1979). Implementing a human services program: How hard will it be? *Public Policy, 27* (4), 385-435.

Cialdini, R. B. (1993). *Influence: The psychology of persuasion* (Rev. ed.). New York: Quill.

Clotfelter, C. T. (1995). Seminar at Duke University (March 8).

Clotfelter, C. T., & Ladd, H. F. (1996). Recognizing and rewarding success in public schools. In H. F. Ladd (Ed.), *Holding schools accountable: Performance based reform in education* (pp. 23-63). Washington, DC: The Brookings Institution.

Drucker, P. R. (1980). The deadly sins in public administration. *Public Administration Review, 40* (2), 103-106.

Dupont, K. P. (1996, September 16). Cup is US'—for now. *The Boston Globe*, p. D2.

Elmore, R. F., Ablemann, C. H., & Fuhrman, S. H. (1996). The new accountability in state education reform: From process to performance. In H. F. Ladd (Ed.), *Holding schools accountable: Performance based reform in education* (pp. 65-98). Washington, DC: The Brookings Institution.

Gould, S. J. (1989). *Wonderful life: The Burgess Shale and the nature of history*. New York: W. W. Norton.

Gould, S. J. (1994). The evolution of life on earth. *Scientific American, 271* (10), 85-91.

Hanushek, E. A. (1996). Comment. In H. F. Ladd (Ed.), *Holding schools accountable: Performance based reform in education* (pp. 128-136). Washington, DC: The Brookings Institution.
Hanushek, E. A., et al. (1994). *Making schools work: Improving performance and controlling costs.* Washington, DC: The Brookings Institution.
Hatry, H. P., Alexander, M., & Fountain, J. R., Jr. (1989). *Service efforts and accomplishments reporting: Its time has come: Elementary and secondary education.* Norwalk, CT: Government Accounting Standards Board.
Hatry, H. P., Fountain, J. R., Jr., Sullivan, J. M., & Krewer, L. (Eds.). (1989). *Service efforts and accomplishments reporting: Its time has come: An overview.* Norwalk, CT: Government Accounting Standards Board.
Herzberg, F. (1968). One more time: How do you motivate employees? *Harvard Business Review, 46* (1), 53-62.
Kahn, A. (1993). *Punished by rewards: The trouble with gold stars, incentive plans, A's, praise, and other bribes.* Boston: Houghton-Mifflin.
Kerr, S. (1975). On the folly of rewarding A, while hoping for B. *Academy of Management Journal, 18* (4), 769-783.
Ladd, H. F. (Ed.). (1996a). *Holding schools accountable: Performance based reform in education.* Washington, DC: The Brookings Institution.
Ladd, H. F. (1996b). Introduction. In H. F. Ladd (Ed.), *Holding schools accountable: Performance based reform in education* (pp. 1-19). Washington, DC: The Brookings Institution.
Maslow, A. H. (1943). A theory of human motivation. *Psychological Review, 50* (4), 370-396.
Meisels, S. J., Dorfman, A., & Steel, D. (1995). Equity and excellence in group administered and performance-based assessments. In M. T. Nettles & A. L. Nettles (Eds.), *Equity and excellence in educational testing and assessment* (pp. 243-261). Boston: Kluwer.
Osborne, D., & Gaebler, T. (1992). *Reinventing government.* Reading, MA: Addison-Wesley.
Peters, T. J., & Waterman, R. H., Jr. (1982). *In search of excellence: Lessons from America's best-run companies.* New York: Harper & Row.
Pillsbury, J. B. (1996). Seminar at Duke University (April 16).
Riordon, W. L. (1963). *Plunkitt of Tammany Hall.* New York: E. P. Dutton.
Shepard, L. A., Flexer, R. J., Hiebvert, E. J., Marion, S. F., Mayfield, V., & Weston, T. J. (1996). Effects of introducing classroom performance assessments on student learning. *Educational Measurement: Issues and Practice, 15* (3), 7-18.
Tully, S. (1994). Why go for stretch targets? *Fortune* (November 14), 145-158.
Wainer, H. (1993). Measurement problems. *Journal of Educational Measurement, 30* (1), 1-21.
Walters, J. (1996). Fad mad. *Governing* (September), 48-52.
Weick, K. E. (1984). Small wins: Redefining the scale of social problems. *American Psychologist, 39* (1), 49-59.
Wolff, R. (Ed.). (1990). *The baseball encyclopedia.* New York: Macmillan.

ACCOUNTABILITY IN LOCAL SCHOOL DISTRICTS:
LEARNING TO DO THE RIGHT THINGS

Richard F. Elmore

THE CHALLENGE TO LOCAL AUTHORITY IN EDUCATION

How do we hold schools accountable for student learning? This question is currently at the center of education policy debates in states and localities across the country. The push to hold schools accountable for student performance has its roots in the political economy of schooling in the United States at the end of the twentieth century—a topic deserving of a paper in itself. Suffice to say that states have, over the past twenty years or so, assumed increasing fiscal responsibility for schools as a consequence of well-intentioned efforts to equalize school expenditures. At the same time there has been a growing recognition of the link between education and economic development at the state and national level. There also has been a significant retreat from involvement in education by the federal government, largely as a consequence of federal budgetary politics and the push for deficit reduction. These changes and others have pushed states into playing a more central role in educational

Advances in Educational Administration, Volume 5, pages 59-82.

ISBN: 0-7623-0081-7

governance, a role that some would argue has always been embedded in the Constitution, but never fully realized because of the powerful hold of local democracy over education.

As education has increased in visibility as a state governmental function, its importance as a fiscal and political issue has driven state elected officials to attach more importance to school performance as an indicator of what state money is buying. Nearly all states have, in the past decade or so, increased the visibility and extent of state wide student testing. Within the last four or five years, a majority of states have begun the process of formulating and implementing state wide standards for student performance, against which the success of schools will ultimately be measured.

Notice that in the previous thumbnail sketch of the push for educational accountability, I have barely mentioned school districts, the traditional locus of authority for governance and administration of public schools. This omission is not an oversight; it reflects the reality of policy making in most states that are addressing the accountability issue at the moment. For the most part, state accountability policies focus mainly on state to school relationships, treating school districts, at best, as intermediate units through which state policy passes on its way to the school. State data for accountability purposes are collected, for the most part, on school level performance. Governance structures are formulated to locate the major accountability for student performance with principals, teachers, and often with school site governance councils that represent professionals, parents, and the community. Rewards and sanctions for good or poor performance are directed largely at schools.

Not surprisingly, local school boards and superintendents view this movement toward more direct state to school relations with a combination of defensiveness, discomfort, and alarm. The most common question I hear from the superintendents and school board members with whom I work is, "What's happening to the tradition of local control in education?" To which I typically reply, "It's up for grabs." In fact, formal constitutional authority for education has always been lodged primarily with states, and local school districts are legally creatures of the state. Authority that the state grants, the state may take away. Current trends in state policy making can be seen as states growing into and claiming their formal role in the governance of education, rather than as intrusions on some mythic state of nature described as local control. States are intervening directly in schools because they see the immediacy and urgency of improving school performance, and they often view school districts as part of the problem rather than part of the solution.

In my work with superintendents and school boards, I urge them to think of the present period *not* as a threat to the tradition of local control, but as a challenge for localities to define a role for themselves that is consistent with the push for greater school autonomy in matters of organization and instruction and greater school accountability for student performance. If localities are seen by state policy makers as defending what they think are their traditional governance prerogatives rather than developing a new, performance based role, they will simply be bypassed by state policy. The political and economic forces that are moving state policy makers toward a more direct relationship with schools around student performance are far more powerful than any traditional arguments for local governance prerogatives that superintendents and board members can raise. My advice to local actors generally is, "get on top of the wave and ride it, or you will probably be washed out to sea."

So, the focus of my analysis in this paper is how to reconstruct the local role in education around the central principle of *adding value* to student performance in schools. Given the current direction of education policy in the United States, local school systems have no choice of whether to embrace this principle; if they are to survive as institutions, they will have to embrace it, or they will be displaced. It often escapes the attention of local administrators and school board members that there is no necessary reason why local educational authorities should exist, at least as independent units of local government. Some industrialized democracies, notably Australia and New Zealand, have simply done away with local educational authorities in their movement toward greater school autonomy and accountability. In other industrialized democracies, notably the United Kingdom, the national government has engaged in policy making specifically designed to undermine the authority of local education agencies. Other industrialized democracies, notably the Scandinavian countries and Germany, treat local education authorities as subsidiary functions of general government at the local level and as instrumentalities for the implementation of state and national policy, rather than as independent governance structures. In these countries, the powerful link for purposes of determining what gets taught to whom and who is accountable for demonstrating that the work is getting done run mainly from the state or national ministry of education to the school, and municipalities are more or less forced to justify their role as productive partners in this relationship. As education becomes more central to the agenda of economic development in the United States and as the performance of American schools comes under increased public scrutiny with the introduction of more and more extensive systems of student performance assessment, local school systems will be forced increasingly to justify their existence in terms of their contribution to what schools produce, rather than in terms of a nostalgic myth of local democracy.

I will focus my analysis on the local district's role in performance based accountability systems, which I will define shortly. I will not be addressing, except in passing, alternative accountability systems, such as market based or contract based systems, although I have much to say about those models. I have limited my focus to performance based accountability systems because alternative systems are well treated by other participants in this symposium and because I think the problems of performance based accountability systems are complex enough to merit separate treatment. My relative lack of attention to alternative systems of accountability should not be read as signaling a belief that they are not important or worthy of attention. Quite the contrary, I think they are very important and are likely to be very visible in policy discussions of accountability in schools.

I will not be reporting new data in this paper. Rather, I will be developing an argument and an analysis of the issues involved in defining a local role in education around the principle of value added to student performance. As part of the research program of the Consortium for Policy Research in Education, my colleagues and I are just beginning a new multi year project on school accountability. I will be drawing on some of the conceptual and exploratory work we have done on that project and putting this work in the context of the accountability problem in local school systems. I will also be drawing on previous work I have done on the role of local school systems in instructional improvement, specifically the example of Community District 2 in New York City, which has developed a unique approach to the related problems of accountability and instructional improvement.

PERFORMANCE ACCOUNTABILITY: DOING THE RIGHT THING AND KNOWING HOW TO DO THE RIGHT THING

The idea of performance based accountability systems has a great deal of currency at the moment in discussions of public sector management. In its basic outlines, the idea is appealing and relatively simple: Public sector organizations should be managed on the basis of performance, rather than process. The central task of public sector managers is to set goals and devise measures of performance based on those goals, rather than to meddle in the day-to-day details of operations in the units that deliver services. Authority for operational details should be devolved to the lowest unit possible in the system. These units should organize themselves into work groups that solve operating problems, and work groups should, in turn, be given authority to change key processes and organizational structures in line with their own knowledge of what works best. The role of system level managers is to set goals, manage the allocation of material and human resources against those goals, devise and administer performance measures, monitor the performance of units, administer rewards and sanctions based on measured performance, and stay out of the way of detailed operating decisions at the unit level. The argument for performance based systems of accountability is that they use the resources of the organization more efficiently, that they are well adapted to functions that require high judgment in the accomplishment of the central tasks of the organization, and that they capitalize on the knowledge and competence of members of the organization by freeing them from the constraints of meddling by higher level managers.

Applied to school systems, performance based accountability stresses the school as the key operating unit of the system and enhancing student academic performance as the main goal of the whole system and of its operating units. Schools should be relatively free to devise their own solutions to the performance problem by, for example, developing specific curricula, teaching methods, student grouping strategies, and ways of organizing the work of teachers consistent with their views of what works to enhance student performance. The role of system level managers is to set goals for student performance, administer student assessment systems that track student performance by school, and administer rewards and sanctions based on measured school performance. Sometimes school level organization is designed to include decision making or advisory bodies composed of education professionals, parents, and community members who are charged with overseeing school operations and bringing the views of stakeholders to bear on internal school decisions--a form of organization commonly called site based management (SBM).[1]

In its simplest form, performance based accountability is a generic model of management. As with any generic management model, it raises difficult and perplexing problems when applied to a specific sector, like education or health care. The fact that the generic model raises difficult problems in the context of specific sectors is not, in itself, a criticism of the model; it is an indication that no general model works well unless it can be adapted to the special problems of a particular kind of organization. Educators have been experimenting with performance based accountability models at least since the nineteenth century, when the British tried something they called "payment by results." In its current incarnation, performance based accountability has taken root in a number of state and local education reforms, but it is highly developed at present in only a few settings--notably in states like South Carolina and Kentucky and in localities like Dallas, Texas.

The early experience of states and localities with performance based accountability systems has received detailed study and analysis.[2] As one might expect, the range of problems emerging from this experience is daunting, but not unexpected. States and localities are wrestling with how to set student performance standards; how to design student assessments that embody the standards they set; how to evaluate school performance; and how to use performance measures, incentives, and sanctions to improve schools and minimize perverse or unanticipated results. These issues are fundamental to the success of performance based accountability systems in education, and there is an enormous amount of learning going on in states and localities around these issues at the moment. Because these issues are relatively well covered in the existing research literature, including my own work with colleagues on state accountability systems, I would like to focus my attention in this paper on an issue that is fundamental to the success of performance based accountability systems but that is not treated well by the existing research literature. I will call this the capacity issue, or the problem of knowing how to do the right thing.

One of the key limitations of the generic model of performance based accountability is its assumption that people working in operating units of the organization have access to the knowledge they need to make the changes necessary to improve their performance as a group. In schools, this means that teachers and administrators know what to do, or can get access to this knowledge through some means, to meet the student performance goals set by state or local authorities. The literature on high performance workplaces and in subfields like Total Quality Management tends to romanticize the energy, competence, and creativity of the workforce in an organization. The problem of mobilizing knowledge to enhance performance is often portrayed in the management literature as a form of liberation and empowerment: People working in organizations are characterized as brimming with ideas for how to improve their work. All managers have to do is to design organizations in a way that liberates these ideas.

My own work on school restructuring, with my colleagues Penelope Peterson and Sarah McCarthey, suggests that this romanticized view of high performance workplaces is fundamentally flawed, at least in its application to school (Elmore, 1996). We studied the relationship between organizational change and classroom teaching practice in a detailed way in three elementary schools. These schools were all engaged in extensive organizational restructuring along the lines envisioned in the literature on high performance organizations. Teachers worked in groups to redesign the curriculum, to reconfigure student grouping structures, and to divide their work with students in new and creative ways. They were empowered to seek help and advice from the outside, and they participated actively in professional development activities explicitly designed to provide them with new knowledge about how to do their work. They did, in other words, virtually everything that people in high performance organizations are supposed to do. Yet, in two of the three schools we studied, all of this organizational restructuring and active participation resulted in little or no change in classroom teaching practice. Teachers continued to teach in much the same way they were teaching before they were empowered, and worse yet, they often *thought* they were teaching differently, even when they weren't. They were highly motivated, they felt good about their work, they were energized, but these changes in the quality of work life had little or no effect on teaching practice and on the experience and performance of students. In the third school, teachers and administrators took a different approach. After a period of experimentation with organizational change, which many regarded as frustrating and unproductive, they focused on a set of ideas about teaching and learning and gradually, over a period of several years, developed an organizational structure that enabled them to implement these ideas with some consistency throughout the school. The organization they evolved was more conventional than that of the other restructured schools, but the results for classroom practice and student learning were dramatic, consistent, and sustained. Rather than focusing on changes in structure designed to empower teachers, they focused on the deliberate development of a common view of teaching and learning and treated the organization as an instrument for accomplishing this view. As in the other schools, teachers were highly motivated and energized by their work, but their work was more productive in the sense that it resulted in relatively consistent changes in teaching and learning over time.

The lesson we took away from this research is that changes in the structure of work in organizations have no direct, apparent, or inevitable relationship to changes in teaching practice or student learning. It sounds obvious, but it is terribly important to say: If you want to change teaching practice and student learning, you have to focus on changing teaching practice and student learning, rather than engaging in largely symbolic structural change around existing teaching practices and existing patterns of student learning. Teaching is hard and demanding work. It involves the most profound conjunction between a deep understanding of content, of pedagogical practice, and of the processes by which students come to understand content. Teachers tend to settle into fairly stable patterns of practice relatively early in their careers and, other things being equal, are often quite unselfconscious about how durable these patterns are because they have virtually no opportunity, in the course of their work, to understand their own practice or to observe and analyze practices that are markedly different from their own. In the absence of sustained analysis of their practice, or observation of other practices, teachers tend, as anyone else would under the same circumstances, to do what they have always done, sometimes relabeling it in response to the prevailing jargon of the day.

The relevance of this work to the problem of performance accountability in schools is that performance standards and organizational change within schools will probably not produce broad scale changes in student learning, unless they are accompanied with strong and direct attention to the problem of changing teaching practice. Teachers, for the most part, will not spontaneously invent new practices that produce new sorts of student learning, even when they work in organizations that are congenial and supportive to the development of new practice. Changing teaching practice is a distinct and complex problem that requires distinct and complex solutions, not just reorganizing around the work. Setting student performance standards communicates public expectations about the right thing to do, an important function of public policy. Changing organization can influence teachers' and administrators' capacities to do the right things. But the fundamental problem for educators is knowing how to do the right things to produce results consistent with public expectations. Most of what states and localities are doing in the name of performance accountability in schools has nothing to do with helping educators know how to do the right things. Setting standards and changing organizational and governance arrangements does not, by itself, address the problem of knowing how to do the right things.

Over the past several years, I have been examining how school districts approach the problem of instructional improvement: how they organize themselves around this function, what they actually do to promote instructional improvement, and how this function fits into the other functions that districts perform. My depressing conclusion is that most districts are ill equipped to affect teaching practice in any sustained, systemic way. Virtually every school district has a curriculum and instruction function, staffed by subject matter specialists, and virtually every district spends some portion of its resources on staff development for teachers and administrators. But in the typical district, these functions tend not to be organized around a coherent view of what good teaching looks like in the classroom. They tend to operate as specialized functions outside the line management relationships between district administrators and school principals, and they tend to use staff development approaches that have weak and scattered influences on teachers working in classrooms.

In the typical district, the instructional improvement function amounts to setting a fixed number of staff development hours for teachers, which typically result in teachers being pulled out of their classrooms to participate in workshops where they listen to a consultant talk about teaching and are provided with materials or sample lessons to take back to their classrooms. Even districts that engage selected teachers in curriculum development projects, often over the summer, tend not to connect these activities in any sustained way to broad scale strategies for reaching large numbers of teachers in their classrooms. The result of this typical district approach to instructional improvement is that teachers experience staff development as a series of disconnected, isolated events that have little relevance to their daily practice, and they feel as they are somehow responsible for figuring out how to translate vague ideas delivered in "talking head" formats into concrete practices—a task they find daunting and largely unproductive. Furthermore, since staff development activities are not integrated into the line management function of the school systems, the signals that teachers receive on a daily basis about what they are responsible for teaching, if any, usually have little to do with the signals they receive from staff development specialists.

My diagnosis of this problem is that it is a classic example of bureaucratic overspecialization. School districts respond to the necessity to address instructional improvement by creating specialized units, staffed by subject matter specialists who have no direct line responsibility for the daily functioning of schools. These specialized units do what they are supposed to do: They run professional development activities for teachers and principals around topics that higher level administrators determine are district wide priorities. But because staff development and instructional improvement are treated as specialized functions, apart from the daily work of teachers and principals, they have only an episodic and indirect influence on what principals and teachers do. Staff development becomes one among many external disturbances in the environment of schools to which principals and teachers have to respond and adapt.

In addition, staff development activities embody theories of learning that are completely inconsistent with the objective of sustained, systemic improvement in teaching practice and student learning. These activities are typically done in isolated workshop settings away from where teachers actually practice. They often involve methods of instruction that are completely inconsistent with the teaching practices they are supposed to be promoting, such as about how to elicit student engagement in learning. They are usually one shot affairs with no systematic follow-up on their effects in the classroom. And, they usually isolate teachers from the administrators who oversee them, so teachers are exhorted to practice in ways which their principals are unfamiliar with and unprepared to practice themselves. What's going on here is not so much the enactment of a faulty theory of adult learning, but the blind pursuit of a set of administrative routines without any reflection of the problem of adult learning.

A handful of school districts have attempted to face this problem head-on by making staff development and instructional improvement a central part of the daily life of schools. I have been studying one such district, Community District 2 in New York City, over the past three years in an attempt to learn what is entailed in addressing the problem of changing teaching practice on a system wide basis in a sustained way (Elmore & Burney, 1996). The District 2 strategy is ambitious and far too complex to describe well in this limited space. But I would like to focus on two dimensions of the strategy that I think are particularly relevant to the problem of performance accountability.

First, for District 2 administrators, school administration *is* staff development; staff development is not a specialized function in the district office, it is what district administrators do. This principle is reflected in the internal management and accountability structure of the system. Principals are responsible for specifying annually what staff development activities they will pursue in their schools, directed at specific content areas, specific grade levels, and specific teachers, and organized around district level priorities for instructional improvement and student learning. Anthony Alvarado, the district superintendent, communicates in his actions, in the organization of the district office, and in the specific activities his staff pursues the core philosophy that "it's about instruction, and only about instruction." Principals and teachers are organized into a variety of crosscutting networks around basic instructional problems in content areas. District wide principals' meetings focus exclusively on problems of instruction and not on the bureaucratic routines and procedures of the district. Principals are expected to participate with teachers in staff development activities around academic content, and they are expected to teach demonstration lessons in classrooms and to be observed doing so by the teachers in their schools. Staff development consultants spend the majority of their time working in classrooms with teachers and only a small fraction of their time delivering workshops outside of classrooms. Teachers are routinely assigned to work with other teachers in their classrooms to learn and observe new practices. Groups of teachers and principals take time away from their daily work to visit classrooms and schools that exemplify successful practices. Aspiring principals are required, as part of the interview process in the district, to visit a series of classrooms and to analyze and explain what they see occurring between teachers and students in those classrooms. There is no distinction in the district office or in schools between the function of staff development and the function of managing schools; they are one and the same thing.

A second noteworthy dimension of the District 2 strategy is that it embodies an explicit theory of adult learning. This theory is that teachers learn to teach differently, first, by understanding and analyzing the limits of their existing practice; second, by engaging in inquiry about and observation of alternative forms of practice; third, by receiving explicit instruction in alternative forms of practice in which the instructional methods of the training are consistent with the instructional methods of the practices that are being taught; fourth, by observing skilled practitioners working in classrooms and by being observed teaching by skilled practitioners; and fifth, by receiving consistent, well-informed feedback from administrators who are themselves capable of practicing in the ways they expect teachers to practice and whose supervisory behavior models the practices they expect of teachers. This turns out to be a particularly powerful theory of adult learning. It is not the only such theory that could be used to address the problem of changing teaching practice. My point is more that the District 2 strategy *has* an explicit theory, in contrast to most other school districts I've observed, and that they are willing to organize their work consistently with that theory.

The District 2 solution to the problem of instructional improvement is a particularly radical one. It is one of many possible solutions to the problem of how to influence teaching practice in a sustained and systemic way. There are many possible solutions to this problem. My intention in using this example is not to proselytize about District 2; it is, rather, to illustrate how school districts have to rethink their conceptions of what it means to manage a school system if they are to be serious about addressing the problem of knowing how to do the right thing. Simply setting student performance standards, engaging in bureaucratic box shuffling, and fussing with novel governance arrangements probably won't produce discernible effects on what teachers teach and what students learn. The central problem of performance based accountability is the hardest part: learning how to reach large numbers of teachers in their classrooms, as they engage in the daily work of teaching, in ways that are designed to provide them with the knowledge they require to teach differently and to produce different results for students.

Hence, I return to the theme of the first section of this paper. The sole criterion for deciding how to organize school districts should be *value added* to student performance. There are, to be sure, some number of central administrative functions that have to be done, or at least contracted for, that do not bear immediately on instruction. But if local school systems are to survive, they have to be organized as if their sole purpose were to promote teaching and learning in classrooms, not in a symbolic way, but in a tangible and concrete way. My hunch is that, if this were the case, central administrative offices would be: (a) much smaller than they are at present, (b) far less specialized in their administrative functions, (c) far more oriented toward direct engagement with principals and teachers around concrete problems of instruction in specific content areas at specific grade levels in specific schools and classrooms, and (d) far more oriented toward mobilizing knowledge about teaching practice and far less oriented toward the procedural regulation of principals and teachers. My hunch also is that districts oriented toward value added to student performance would: (a) hire teachers, principals, and central office administrators based on their knowledge and skill in dealing with fundamental problems of instruction; (b) evaluate all administrators in terms of their explicit contribution to the improvement of instruction; (c) redesign the work of the organization to focus discretionary resources on the creation of opportunities to learn new practices at all levels of organization; and (d) as learning progresses, formulate increasingly explicit expectations or standards for what good teaching and administrative practices look like and what students, teachers, and administrators are expected to know and be able to do in the normal performance of their daily work.

Educational systems should have as their central focus enabling the learning of everyone who works in them. If this were the case, the problem of performance based accountability would be much less daunting, much more engaging, and much more understandable than it currently appears to be. People working in school systems would routinely learn to gauge their daily actions based on whether they are contributing to someone else's capacity to do the central work of the organization. Reorganizations, meeting schedules, governance issues, and personnel decisions would be routinely cast as decisions about enabling good classroom practice. Hence, learning how to do the right thing would be seen as the work of the organization, rather than as a specialized administrative function performed by people with no administrative authority over schools.

I have had the opportunity to talk with large numbers of school professionals about these ideas over the past two years, through our summer continuing education programs at the Harvard Graduate School of Education: notably, the Harvard Institute for School Leadership, a team based, multirole program focused on systemic school improvement, and, more recently, the Harvard Institute for New Superintendents, aimed at superintendents who have taken their positions within the last year. School professionals have a variety of reactions to these ideas. Overall, my unsystematic reading of responses suggests that peoples' enthusiasm for the ideas is roughly inverse to the square of the distance from the classroom in their present position. That is, enthusiasm wanes rather sharply as one moves from teachers to principals to central office administrators to superintendents. Most teachers regard it as obvious, although seldom the case, that principals should be engaged in learning the same practices that they, as teachers, are being asked to learn. Principals are usually quick to embrace the idea that their job entails enabling their own learning and the learning of teachers until they are confronted with a tough choice about resource allocation or the use of their own time. Then, they tend to retreat into what they describe as the "realities of the job."

District administrators are generally sympathetic to the idea that their work should be evaluated on the basis of how well they contribute to other peoples' learning, until they are faced with a problem that requires them to deal with an angry parent whose convictions are at odds with their own or until they face the prospect that one consequence of this principle might be that their job may be redefined or go away altogether. Superintendents often respond to this argument like the proverbial deer in the headlights—sort of a stunned, "what-have-I-gotten-myself into-now" look, followed by elaborate attempts to educate me in the realities of the modern school superintendency. I am led by these interactions to conclude that most people who work outside the classroom in school systems have only very sketchy ideas about how their work contributes concretely to what a given teacher is doing with a given group of students at a given time on a given day with a given subject. Given what is happening in the policy environment of public education at the moment, I don't think this sketchiness bodes well for the future health of school districts.

I would like to focus on three specific responses that experienced educators make to the argument that their work should be judged solely by its contribution to the learning of teachers and students because I think they illustrate the magnitude of the task that local school districts face in finding a role for themselves in a world of performance based accountability. First, one particular superintendent responded to my argument by saying roughly the following:

> I don't see my job as having much to do directly with teaching and learning. In fact, my job is really to do the things that protect principals and teachers from

> the distractions that keep them from focusing on teaching and learning and to take care of the routine administrative functions that enable schools to function smoothly.

Another superintendent said,

> I know enough about this job now to know that the problems the world throws at you don't have much to do with the problems of teaching and learning. People complain about the bus schedule, about the football coach, about the graffiti in the parking lot, and someone has to deal with these issues. I end up doing a lot of that, not because it's what I want to do, but because someone has to do it. So I think the realities of the job are that there are an infinite number of distractions from the central work of the system.

I have heard various versions of these arguments dozens of times in a number of settings where I work with superintendents and school board members. I do not take issue with the accuracy of these portrayals of the realities of the modern school superintendency. Public schools do face complex and daunting demands from their constituencies, and these demands often have nothing to do with the core functions of teaching and learning. What puzzles me about these arguments is their passivity. They portray the chief executive officer of a large public enterprise as someone who has no influence, or precious little influence, over the core processes of the enterprise—indeed, in some formulations, they portray the superintendent as someone whose role is explicitly to *avoid* any direct influence over the core processes of the enterprise. Would we accept an explanation from a hospital administrator, or a chief medical officer in a hospital, in which they said they really had no direct influence on mortality rates in the hospital's surgery department because their real job was to deal with the complexities of health insurance, food service, and clean linens so the doctors could get on with their work? I doubt it. No one disputes that health insurance, food service, and clean linens are instrumental in some way to mortality rates in surgery, but if the chief administrative officers of a hospital don't, in some material way, take direct responsibility for how well medicine is practiced in various parts of the organization and whether practitioners have the knowledge and resources they need to practice effectively, then they will almost certainly be ex-administrators.

Using the complexities of the organization and its environment to explain away the responsibility of administrators for the organization's core functions is an almost certain invitation simply to remove administrative responsibility and give it to someone who will assume it. As pressure builds for performance accountability in schools, superintendents who use the complexities of the job as a shield from responsibility for teaching and learning will probably be ex-superintendents, or alternatively, the systems they used to administer won't be there anymore.

Another common response I often hear to the argument that school systems should take direct responsibility for the improvement of teaching and learning is that good teaching is such an uncertain, precious, fragile phenomenon that it cannot be managed in the way other services are managed. It is the job of administrators to liberate good teachers and get out of their way, the argument goes, not to try to directly influence how teaching and learning occur in classrooms. One mixed group of teachers and administrators, for example, reacted to the District 2 example with mixed enthusiasm. Some members of the group argued that the District 2 strategy was too heavy-handed and prescriptive, especially when they discovered that principals had actually lost their jobs as a consequence of not meeting expectations for teacher development and that something like half the teaching force in District 2 had turned over in the eight years or so since the district had embarked on its instructional improvement strategy. It was good for administrators to feel like they should know something about teaching and learning, they argued, but they were extremely uncomfortable that anyone should be making judgments about what constituted good teaching, especially if such judgments might result in people losing their jobs.

At the root of this perspective is what I would call the "trait theory of teaching." In this view, good teaching is something that descends out of the sky into teachers, or rises up out of their genetic material. Good teaching is a mysterious gift that no one can explain, and it manifests itself in individual teachers as a trait, much like height or shoe size, that is not susceptible to alteration once it is fixed. Hence, one can't really organize or manage for good teaching; the basic reality is that some teachers are good and some aren't. Furthermore, since good teaching is a trait, one can't really make judgments about it, or hold anyone responsible for it, any more than one would hold someone responsible for height or shoe size.

To say that teaching is a complex, uncertain, and highly skilled occupation is most definitely *not* to say that it unsusceptible to judgments of quality. If this were true, we would be unwilling to say that there is no way of judging the quality of a lawyer, a psychiatrist, an actor, an architect, or an aircraft designer. In fact, society routinely makes judgments about how well practitioners of these occupations do their work, and people more or less routinely lose their jobs as a consequence of not being good enough to meet a reasonably rigorous standard of performance. We don't say that bad practitioners of these occupations are bad people because they can't meet our standards of quality, or that they are defective in their genetic make up; we simply say that they ought to find something else to do in this life. Maybe their incompetence has something to do with their aptitude for the work, but the basic point is that we focus on their demonstrated capacity to do the work in deciding whether they should be able to practice the occupation. Education is one of the few remaining occupations in our society where one still routinely hears that it is impossible to make judgments of quality because the work is complex. As pressure for performance accountability in schools, this argument will become increasingly less plausible.

A third response I often hear to the argument that school systems should focus on teaching and learning is that educational administrators are simply unprepared to do the hard work of making difficult judgments about the quality of teaching and learning. Gifted teachers often say something like the following:

> I would love to work with a principal who is interested in helping me improve my teaching, but the basic reality is that I haven't met many who actually have anything useful to contribute to my teaching. I'm not about to subject myself to the judgments of someone who hasn't been in the classroom in so long he can't remember what it's about, who probably got his job by doing something that has no relationship to teaching, and who doesn't spend enough time in my classroom to understand what I'm trying to do.

Comments like this are, I think, symptomatic of life in organizations that have lost track of their core functions. If it is actually true that school systems select and promote people to principalships who have neither the competence nor the authority to engage teachers in discussions of teaching practice and to make judgments about what they see happening in classrooms, then the systems are simply not designed to do what they are supposed to do. If they are not designed to do what they are supposed to do, then the systems will fail, and the people who work in them will fail along with the systems.

We can prolong the agony of failure by developing elaborate explanations for how we got to where we are, but the basic reality is that organizations that don't pay attention to the improvement of their core processes ultimately fail. In the case of schools, they fail in increasingly conspicuous ways, and in the process of failing, they often engage in the unseemly practice of blaming their clients, society-at-large, and dark impersonal forces over which they have no control for their failure. At some point, though, it becomes obvious that a big piece of the failure lies within the organizations themselves and the people who work in them, and society simply finds a way to get the work done without relying on failing organizations. If local school systems don't select, promote, and cultivate the talent of people who understand how to nurture continuous improvement of teaching and learning, then the teaching and learning function will simply migrate to some other set of social institutions.

Performance accountability in schools, then, is only partly about constructing policies and accountability systems that tell schools what they should be producing by way of student performance—knowing the right thing to do. The deeper, more fundamental problem is getting schools to come to terms with the processes by which educators learn *how to do the right things* to produce the kind of student learning that society expects. Local school systems can play an important role in solving this problem by judging everything they do in terms of the criterion of value added to student performance. But the gap between where school systems need to be to do this job well and where most school systems currently are is huge, and the environment of school systems is becoming increasingly inhospitable and increasingly unconvinced by the explanations that school people offer for why they are not able to do what society expects of them.

SOME PRINCIPLES FOR DESIGN AND ACTION

My first rule of teaching, which I learned from watching kindergarten teachers, is "never make a mess any larger than you can clean up by the end of class." I am afraid I have made a rather large mess here, and I feel some obligation to tidy it up around the edges, if not clean it up entirely. I have tried to say as strongly as I know how that the economic and political pressures driving the performance accountability movement in education are large and impersonal forces that are indifferent to the future of local school systems. I have also made it clear that I think local school systems are, for the most part, poorly prepared to respond to the challenges presented by performance accountability systems and that the future of local school systems depends heavily on their capacity to learn, as quickly as possible, how to improve their core processes of teaching and learning in a deep and sustained way around the criterion of value added to student performance. I am not prepared to give detailed prescriptions for what school systems should do, in part because I am still working on this problem in my own research and in part because I think there are many possible solutions to the problems I have presented here that will work differently in different settings, and I don't pretend to be able to prescribe for all circumstances. But I do think we know enough about the basic shape of the performance accountability problem and its possible solutions to sketch out some principles that can inform the design of local accountability systems and the actions that educators can take to make progress on the problem. So I will conclude by laying out some of these principles. *The Roots of Accountability: Responsibility and Expectations*[3]

I mentioned earlier that my colleagues and I at the Consortium for Policy Research in Education are beginning a multi-year study of school accountability. As part of this project, we have had to construct a simple working theory of accountability in order to decide such mundane things as what kind of schools to study and what questions to ask when we are interviewing people. One of the most powerful insights I have discovered from reading the literature on accountability and formulating a working theory is that the idea of holding someone accountable rests on certain prior conditions. Briefly, accountability means, literally, to give an account of one's actions to an external authority. In the case of performance accountability in schools, it means giving an account of one's impact on student learning as a consequence of the public money one has spent. But formal accountability, in this sense, works in concert with other powerful normative systems in schools. At the individual level, we call these systems *responsibility*; at the collective level, we call them *expectations*. The ability to account for one's actions depends, we argue, on the alignment of responsibility, expectations, and formal accountability. Let me illustrate.

Suppose I am a ninth grade algebra teacher in an urban school system that has recently decided that all students in the system should take algebra and that schools will be evaluated, rewarded, and sanctioned based, in part, on how well students learn algebra. Now, figuring out how to teach algebra to large numbers of students who would not have previously taken the subject is a formidable task. Simply measuring how many students have taken the subject and how well they do on a test of algebra knowledge—a basic element of any formal accountability system—will not necessarily result in students learning algebra. Somehow I, as a ninth grade algebra teacher, need to internalize the idea that I, as an individual, am responsible in some important way for figuring out how to teach algebra to students who I might not have taught before. If I don't take some level of personal responsibility for this task, the likelihood that you will coerce me into being accountable for results on an algebra test, regardless of how you try to do it, is very remote. I can blame the students for not learning the subject. I can blame their families for not reinforcing what I am trying to do in the classroom. I can blame other teachers for not teaching the prerequisite math skills. And, I can blame you for creating a stupid accountability system that doesn't take account of how difficult my job is. So, a fundamental condition of your accountability system working is that I am prepared to take responsibility for actually figuring out how to teach algebra.

Even if I take responsibility, I am only one teacher in a large system. To some degree, depending on the specific design of the accountability system, I am being held accountable not just for my own teaching but also for my colleagues' teaching and, perhaps most importantly, for students being willing to engage in the difficult task of learning algebra. Holding me accountable, therefore, is not just a problem of my taking responsibility, it is also a problem of group expectations. I have to be able to plausibly expect that my colleagues in the math department and in earlier grades will take the same view of the importance of algebra as I do, and more importantly, that students will accept the expectations of their teachers as legitimate enough to engage in the effort of learning. No matter how responsible I am, my teaching will have only a limited impact if my responsibility isn't translated into mutually agreed upon and enforceable expectations.

The social connective tissue that makes formal accountability possible, then, is a certain degree of consistency and alignment between individual responsibility, collective expectations, and the requirements of formal accountability systems. The practical consequence of this working theory is that one should not expect formal accountability systems to work if they are not consistent with the underlying social norms of responsibility and mutual expectations. A big part of the task of holding schools accountable for performance at levels they have never achieved before is convincing teachers and students to recalibrate their existing conceptions of responsibility and their existing expectations of each other. How does this happen?

EXISTENCE PROOFS AND THE THEORY OF SMALL TRIES

Al Shanker was fond of saying that much of the argument around performance standards and accountability in education is analogous to trying to teach someone how to play basketball by having them sit through hours of lectures about basketball—the theory of basketball, the history of basketball, basketball as a social experience, famous basketball players, and so on. To extend the algebra example, I could be shown curriculum standards that illustrate what I am supposed to teach students about algebra. I could be shown sample examination questions of the types of algebra questions my students will be expected to answer on a performance test. I could even be shown video tapes of teachers teaching algebra in novel ways designed to appeal to students. But, in the final analysis, it is unlikely that I will adjust my conception of my responsibility for teaching algebra or my expectations of my students and colleagues unless I actually have the experience of teaching some aspect of algebra to students I may not have taught before and most importantly, of *succeeding* at doing it. The most powerful influence on my sense of responsibility and my expectations of others, then, is not likely to be you haranguing me about the importance of algebra or even showing me what good algebra teaching looks like. The most powerful influence is likely to be my actually doing it, or some bite sized piece of it, and succeeding at the doing of it. Furthermore, the most powerful influence on my students' and colleagues' expectations is not likely to be someone telling them what to do but actually doing something in which we all succeed, more or less, after a while.

If accountability is fundamentally about individual responsibility and collective expectations, then it makes sense to understand how these basic building blocks are formed and influenced through the concrete actions of students and teachers working together. Telling me, "*Teach algebra*!" is too much, even when you can show me the exam that my students will have to take. Asking me to introduce students to the concept of an unknown and to work through this problem with a group of students and a group of teachers who are all trying to figure how to make the inscrutable subject of what x means clear to each other—now that may get me interested in discovering what might work. Show me that it is possible in a social context where I can express my ideas about it and demonstrate what I can do, give me an existence proof in a context where failing is not catastrophic, and you improve my chances of actually believing that I can do what you are going to hold me accountable for doing.

Changing conceptions of responsibility and expectations, then, is social process. This process can be informed by external performance standards, examination systems, curriculum guidance, and all the other accouterments of formal accountability systems. Indeed, I would argue that formal accountability systems are a key source of leverage on teachers', students', and parents' conceptions of responsibility and expectations. But external accountability systems don't cause people to behave differently. What causes people to behave differently is actually doing something and, after a fashion, succeeding at the doing of it. Hence, building accountability for high standards of academic performance requires a sophisticated understanding of how to construct existence proofs of success. An existence proof is simply a tangible example of how we have succeeded at doing something that we thought we couldn't do, or that we felt unprepared to do, or that we felt highly uncertain about doing. Putting a string of existence proofs together into a coherent plan of action requires developing a theory of small tries: If we succeed at understanding this slippery idea of an unknown, then we surely ought to be able to put x into an equation like $x + 4 = y$ and begin to explore, through graphs and tables, what happens to y when we substitute a number of values of x. If we can do this, then we surely ought to be able to make the equation more complicated and do the same thing. In other words, we affect our sense of what we are responsible for and what we can expect of each other by demonstrating to ourselves that we can do things we previously didn't think we could do.

RECIPROCAL ACCOUNTABILITY FOR CAPACITY AND PERFORMANCE

All forms of accountability are implicit contracts, and the basis of a contract is the principle of reciprocity or exchange. Unfortunately, most designers of formal accountability systems have only a foggy notion of this basic principle. If you ask me to teach algebra to a cross-section of students I have not taught before, I am inclined to ask, "What are you going to do for me to make this possible?" Most often, the answer to this question is, "It's your job to do what I tell you to do; here's an algebra textbook; go do it!" This kind of relationship is hardly a basis for reciprocity.

Let me propose the simple, radical principle that every demand for increased performance through a formal accountability system should carry an equal reciprocal obligation on the part of the party making that demand to provide the capacity to meet the demand. If you demand that I teach algebra to students I have not taught before, then your demand is only enforceable if you can demonstrate that you have provided me with the opportunity to acquire the knowledge and skill necessary to meet that demand. If you are to hold me accountable for performance, I am entitled to hold you accountable for capacity. If this principle were applied to most existing formal accountability systems, I think there would a remarkable drop in the testosterone level of policy makers and administrators. Or, in the language of one primary grade teacher I talked to recently about the problem of accountability, "Just once, I'd like the guys in the suits to come into my classroom and get down on their hands and knees with me and see what it's like to actually teach this stuff. Then, I think We could have an intelligent conversation about accountability."

If the principle of reciprocal accountability for capacity and performance were applied to the organization of most local school systems, I think there would be some fairly radical changes. Administrators and board members would be obliged to say what they were prepared to do to create the capacity to meet the demands they lay on teachers and principals. Central office staff would be obliged to justify their jobs in terms of the net value they add to the capacity of teachers to do their work. Principals would be obliged to say what they were prepared to do to remedy the failures of performance they observe in classrooms. And so on. Most teachers I talk to are suspicious of performance accountability systems, not because they don't believe that they are ultimately accountable for what students learn, but because they believe (justifiably, I think) that most administrators and policy makers don't have any idea of the practical demands they are imposing on teachers and students through their actions. The more sophisticated teachers I talk to think of performance accountability systems as largely symbolic, political gestures directed at the public, not as serious efforts to improve teaching and learning. Regardless of whether these perceptions are accurate, they constitute disturbing evidence of a colossal failure of reciprocity in school systems. Teachers often feel they are being pushed, with their students, further and further out on a long limb of accountability, which at some point will be sawed off, while policy makers and administrators remain safely perched near the trunk of the tree. A more positive metaphor would be something like policy makers and administrators gently passing lumber out on the limb for the purpose of building a tree house that would help teachers and students survive the next windstorm of public scrutiny.

One aspect of District 2's strategy of instructional improvement that I find fascinating is how it embodies this principle of reciprocity. In the daily life of District 2, the work of principals and teachers is given external scrutiny and evaluation that most school people would regard as frightening and intrusive. It is not unusual, for example, for the deputy superintendent—or the superintendent, for that matter—to visit a classroom and observe a teacher, accompanied by the principal, and for the parties to the visit to have a brisk and frank hallway conversation about the teaching they observe in the classroom. This sort of scrutiny would make most educators break out in hives. But in District 2 it is seen as a more or less routine way of doing business. It took me a while, and at least two drafts of my paper on District 2, to figure out what was actually going on in these visits. Then it began to dawn on me that the currency of exchange in these visits was not administrative control, but assistance. That is, central administrators, in some sense, had earned the right to visit, examine, and make judgments about classrooms because they were obliged to offer solutions, or at least to participate in developing solutions, to any problem they identified. They were also obliged to find the resources necessary to implement the solutions they offered. Their credibility and authority as administrators depends on their ability to deliver capacity.

CONTEXT MATTERS

The push for performance accountability is fundamentally a push for higher standards in education. And, the push for standards is, inevitably, a push for some sort of standardization, in the sense of clearer expectations for what teachers will teach, what students will learn, and how much variability we can tolerate in schools around matters of teaching and learning. Standardization is a bad word among educators, because it smacks of bureaucratic control and mindless regulation. But whether you call it standardization or something else, we are interested in assuring that all students receive some high quality version of instruction that is designed to help them meet high expectations for performance. If the result of increased performance accountability were increased variability in student performance (a distinct possibility, but this is the subject of another paper), we would probably regard our efforts as having failed.

On the other hand, schools vary significantly on almost any dimension relevant to their capacity to meet performance standards—in the backgrounds and capacities of their teachers, in the socioeconomic, cultural, and linguistic background of their students, in their relations with parents and community agencies around the welfare of children, and so forth. One of the reasons that performance accountability has so little credibility with classroom teachers is that they feel that accountability systems pay virtually no attention to the special conditions they face in the classroom, dictated largely by the social context of the school. If the success of external accountability systems hinges heavily on teachers taking responsibility for student learning and developing powerful expectations among students and other teachers, then the credibility of such systems hinges heavily on their capacity to respond to variations in school context. Let me return briefly to the algebra example.

I have now internalized the idea that I am responsible, in some sense, for teaching algebra to students I have not traditionally taught. I have also worked with students and other teachers to develop a number of existence proofs and small tries that demonstrate that, together, we can probably figure out a way to make algebra happen in the minds and actions of teachers and students. In the process of doing this, however, we discover that a major constraint on our ability to make algebra happen in our school is that a significant proportion of our students have difficulty reading the textbook because it is written in textbook English and our students, because of their cultural background, are not exactly fluent in textbook English. Now, across town is another junior high school where the problem is not so much decoding textbook English as it is competing with television for homework time. The students in this other school will probably look better initially on the algebra examination because they have less trouble reading the textbook. Do we get penalized in the new accountability system because we have discovered that we have to teach reading at the same time we teach algebra? We should probably be rewarded for discovering that reading is a constraint on our ability to teach algebra. If the accountability system is administered in a ham handed way, however, we will probably be penalized for this discovery.

This is only one example among many that we could devise to illustrate the interaction of context with accountability. But it serves to illustrate a basic principle about making accountability systems work: That teachers and principals should be rewarded for working on the right problems, in addition to be rewarded for learning to do the right thing and for producing the right results. Again, this principle imposes a heavy burden on local school systems if they are serious about the criterion of value added to student performance. The exact combination of actions that adds value to student performance will vary significantly from school to school, and it is the job of local policy makers and administrators to discern the difference between contextual conditions that constrain performance and excuses for poor performance. The credibility of external accountability systems, from the perspective of teachers and principals, depends heavily on a delicate balance between attention to context and attention to system wide objectives. The solution to this problem lies in putting the problems that performance standards raise at the center of system level administrators' concerns, and active problem solving at the center of system level administrators' work. If system level administrators are seen to be bureaucrats administering general standards that have little relevance to particular schools and their problems, then students, teachers, and principals will soon conclude that the purpose of accountability systems is to embarrass and humiliate them, rather than to help them do better. Embarrassment and humiliation is not a particularly robust source of human motivation.

The designers of accountability systems have yet to face this problem squarely. Although there is a lively debate going on right now about whether schools should be evaluated on the basis of their absolute performance, their performance controlling for student background, or the value they add to student learning, I don't think this technical measurement issue is the most important problem. The central problem is rather what school systems are prepared to do to support schools as they adapt general standards to the problems of their particular contexts. Without this kind of support, it won't matter how we measure and evaluate student performance, because schools will be unable to meet any standard of performance we choose.

REPRISE: THE ROLE OF LOCAL SCHOOL SYSTEMS IN PERFORMANCE ACCOUNTABILITY

I have tried to make the case that the debate over performance accountability in public schools is symptomatic of a fundamental questioning of the role of local school systems in the governance and administration of education. The forces that are producing this questioning are, I have argued, indifferent to the preservation of local school systems. And, the long-term survival of local school systems depends on their developing a new role and a new form of organization based on the criterion of value added to student performance. I have tried to illustrate that this criterion has profound implications for the kind of administrators we need to run local school systems, for the role of local school boards and system administrators, and for the relationship between school systems and schools. I have tried also to resolve this analysis into a few tentative principles for design and action.

Educators and school board members are fond of saying that they do what they do "for the kids." I do not doubt that most people involved in public education are motivated by the ideal of helping children learn and that most have made considerable personal sacrifices to do what they are doing. The problem facing public educators, though, is that they are widely perceived as not having acted skillfully enough on their good intentions—that the systems they have created to manage schools have lost track of their central purpose and their core functions and, in many instances, have become powerful engines of student academic failure. The stakes are high in the present debate about performance accountability because they involve not only the success or failure of specific reforms, but also the success or failure of the institutions that govern and manage public education.

NOTES

1. See, for example, Mohrman and Lawler (1996).
2. See, for example, Ladd (1996).

3. I am indebted in this section to my colleagues, Charles Ableman and Susan Kenyon of CPRE-Harvard, and Martin Carnoy and Jennifer O'Day of CPRE-Stanford, upon whose work this formulation draws. I take full responsibility, however, for this formulation.

REFERENCES

Elmore, R., & Burney, D. (1996). *Staff development and instructional improvement in Community District #2, New York City.* Paper prepared for the National Commission on Teaching and America's Future.

Elmore, R., Peterson, P., & McCarthey, S. (1996). *Restructuring in the classroom.* San Francisco, CA: Jossey-Bass.

Ladd, H. (Ed.). (1996). *Holding schools accountable.* Washington, DC: The Brookings Institution.

Mohrman, S., & Lawler, E. (1996). Motivation for school reform. In S. Fuhrman & S. O'Day (Eds.), *Rewards and reform: Creating educational incentives that work* (pp. 115-143). San Francisco: Jossey-Bass.

MOTIVATING EDUCATORS TO DO THE RIGHT THINGS

Robert C. Hill

LOOKING IN FROM THE INSIDE

Defining the roles of members of the board of education and central office administrative staff relative to the successful day-to-day operations of a multi-unit school system has represented a challenge to me from the first moment I became a school superintendent. The path I followed to the superintendency afforded me the opportunity to view school leadership from a variety of perspectives, so that by the time I assumed my current duties I already had begun to be uneasy about one challenging question. My dilemma has been determining how individuals so far removed from the essential tasks of school can formulate policy and steward resources in ways that actually contribute to improved teaching and learning. Just over five years in the superintendency has served only to heighten my anxieties.

Against that backdrop, I offer my thoughts and reactions to the works of Robert D. Behn and Richard F. Elmore presented to the First Edwin O'Leary Symposium on Financial Management. The topics of their papers, including measurement, motivation, accountability, and increasing the capacities of

Advances in Educational Administration, Volume 5, pages 83-92.

ISBN: 0-7623-0081-7

chools to improve teaching and learning, provide a meaningful context for contemplating answers to the question posed. It is my intent to offer the views of a practicing and admittedly perplexed school superintendent in a way that I hope will add to the important dialogue generated at this symposium.

Reflecting on the motivation of educators seems like a good place to start. Slightly more than a year ago, I had the opportunity to be a part of a conversation with Richard Elmore about how to create schools that truly do work—where students learn, where parents are engaged and satisfied, where community support exists, and where the educators feel a sense of pride and accomplishment in what they do. In the course of that conversation, Elmore posed the question, "What really motivates teachers?" I and my colleagues offered a number of thoughtful answers, but none was truly on target with what Elmore had in mind. Eventually, he shared with us the conclusions supported by a broad body of research indicating that teachers are motivated most by the learning success of their students.

Although this point about how teachers are motivated is not directly stated in Elmore's symposium paper, there is little doubt that it does form a strong basis for that paper and that it serves as a powerful connection between his work and that of Robert Behn. I also submit that the knowledge and belief of truth in the idea that teachers are motivated by student learning is central to the uneasiness to which I have admitted in my introduction. Because I have observed a decreasing sense of accomplishment among teachers for much of the past two decades and a weakening of motivation in many cases, I believe I can justify the conclusion that educators recognize the fact that the capacity for high quality teaching and student learning has diminished during this same period of time.

In the comments that follow, I have chosen to first analyze Behn's work and then Elmore's. In several instances I have been able to offer analyses that link parts of the two papers where topic commonalties exist.

LINKING MEASUREMENT AND MOTIVATION

Overall, I found Behn's paper to be meaningful from the perspective of an administrative practitioner. Most of my analysis focuses on two important sections of the paper, one listing the problems of linking measurement and motivation and the other offering suggestions for designing a system that provides for good linkages.

Behn lists the following problems to linking measurement with motivation:

1. Establishing measures is controversial.
2. It is often impossible to measure outcomes.
3. Measures can distort behavior.

4. Traditional holding-people-accountable systems create many more losers than winners.
5. Measures can discourage cooperation.
6. Simple measures can be unfair.
7. Measuring today's status tells us where we are, not how much progress we have made.
8. Complex measures do not motivate anything.
9. Measures encourage the hoarding of limited resources.
10. Measures themselves cannot improve performance; organizations also need resources.

This list led me to ponder two evaluative questions: (1) Are there contradictions between items on the list and perceptions of the experiences I have had as an educator? (2) Is the list comprehensive, or can I contribute to its improvement by offering additional views on why it is problematic to link accountability measures to motivation?

My experience as a teacher, principal, curriculum director, and superintendent all point to a clear validation of the items on the list as they apply to schools and school districts. On the basis of my experience, I cannot refute the points Behn made. I could, in fact, easily offer specific, recent examples of occurrences in the school district where I work to validate each point. I will avoid the temptation to do so, but I will offer several examples that may verify the collective premise of the list.

Two items on the list appear at first to be in contradiction, although they are not. One point offers the view that simple measures can be unfair, and a second point suggests that complex measures do not motivate anything. The apparent conflict raised by these two points actually provides an accurate description of the balancing act between standards for measuring school and student performance and standards that educators face every day. A view of this conflict also provides evidence to support Behn's first premise that establishing measures is controversial.

Most educators believe that teaching and learning is an extremely complex relationship, and they often express frustration that accountability measures imposed by boards of education or state legislatures focus too narrowly and miss important outcomes because of the tendency of those decision makers to simplify complex issues. American educators themselves also contribute to the tug of war of standard setting, as evidenced by their reluctance to embrace any one set of standards, despite the existence of hundreds from which to choose and the countless hours that have been spent developing them and deliberating their merits. When public views are added to the mix, battle lines almost always are drawn.

It is my view that in all of the many public policy debates occurring in the United States today, education offers strong evidence supporting Behn's point about the dichotomy of competing measurement forces. All one need do is witness the warfare that has erupted at local, state, and national levels over standards and the outcomes that accountability systems are intended to measure. The very word *outcomes* now has a political charge to it that has, in effect, erased the term from the practitioners' public vocabulary. Even the teaching of reading has taken on a political edge in the debate of phonics instruction versus whole language approaches. In large part, the disagreement over measuring outcomes and the politics of reading are associated with differing perspectives on simplicity versus complexity in the measurement of teaching and learning. Unfortunately, the public debate of these issues more often becomes a test of who is right and who is wrong, rather than informed discussion of the competing forces of simplicity and complexity in attempts to measure and motivate.

A second major theme in Behn's list focuses on the general issue of resources and how they are affected by attempts to measure and motivate. Behn suggests that accountability systems can produce more winners than losers, discourage cooperation, and lead to the hoarding of resources. I offer a very specific example from my workplace to confirm these ideas.

Several years ago we initiated Program Improvement Grants in the Springfield Public Schools. At the time, the school system was in the process of implementing a move toward site based decision making. Program Improvement Grants were intended to provide an incentive for school improvement teams by making resources available to them to support one or more of their initiatives. The money offered in the grants could be used for the acquisition of materials, for hiring consultants to conduct inservice training, or for buying teacher time to do the work of the school improvement team. At the time, a number of schools had succeeded in implementing some successful instructional approaches, and it was hoped that the grants would provide one mechanism for the dissemination of good practices within the district.

The initial program budget was $400,000. At that level, each school improvement team could have access to a modest grant if it submitted an appropriate application outlining intended uses of the funds. Before the program was initiated, however, the program budget was reduced to $150,000, when $250,000 was moved to cover the costs of bargained agreements with the district employees. That transfer occurred because of the inability of the superintendent to convince the board of education of the grant program's importance.

For two years the grant program operated on a competitive basis. Approximately two-thirds of the district's schools submitted grant applications. Approximately one-half of the proposals were funded. Consequently, one-third of the schools were grant winners, one-third were losers, and one-third were nonparticipants in the process. During that two-year period, not one grant application proposed to use the resources to import instructional practices from another school in the district. Principals expressed strong opposition to the competitive nature of the grant program and the demotivating impact of making comparisons of test scores among schools a means for verifying the existence of best practices.

During the summer of 1996, principals were empowered to alter the program as they saw fit. That fall, Program Improvement Grants were awarded to approximately one-third of the district's schools by a lottery. The amount each school received was based on a mini version of the school aide formula. Each winning school was required to submit a plan for how to use the resources and, pending approval, each had the choice of spending the resources over a one-, two-, or three-year period. The money was distributed on a rotating three-year cycle, until resources might become available to fully fund the program on a year-to-year basis. Is the revised Program Improvement Grant approach better than its predecessor? Only time will tell. What led to the revision? The first system was viewed as demotivating, rather than motivating.

Before discussing Behn's suggestions for creating effective systems to link measurement and motivation, I offer two suggestions for inclusion on his problems list. The first relates to Behn's citation of the research of the Government Accounting Standards Board (GASB) relative to the power of service efforts and accomplishments (SEA) data. A conclusion found in the GASB report suggests six reasons why SEA data should be used more extensively in determining the accountability of government programs. I have no quarrel with the validity of the reasons advanced for making better use of SEA data, and I believe the points made are applicable to schools and school districts. My concern, however, is that in the education policy making that I have observed at the state level in Illinois, I see little or no evidence that elected officials give any serious consideration to the types of data generated in the SEA process. Consequently, I suggest that one additional problem to add to Behn's list is that policy makers frequently ignore the evaluative information generated about existing programs as they consider adjustments to them.

My second suggestion for addition to Behn's list concerns the misconception that data and information are one and the same. Behn accurately notes that education is a data rich endeavor. My experiences, however, suggest that the conversion of data to the broader context of information on which to base evaluative decisions often falls far short of the intended mark. Behn might argue that the reasons for this phenomenon are embodied in several of the problems he included in his list: the need to simplify, the difficulty of measuring outcomes, and the whole issue of complexity. I believe, however, that the confusion between data and information is an important enough phenomenon to deserve its own place on the list. It appears that such confusion often contributes to poor policy decisions, which I believe become very demotivating to the practitioners who are affected by them.

Another major section of Behn's work is summarized in the following list of recommendations for designing a system to link measurement to motivation. The list is comprehensive and potentially useful to school personnel:

1. Start with a few easy victories.
2. Choose simple measures that will motivate the behavior needed next.
3. Give everyone a chance to win.
4. Reward progress, not absolute results.
5. Set achievable "stretch" goals.
6. Do not rely upon the same measure to motivate all schools.
7. To motivate team behavior, reward team behavior.
8. Do not rely upon monetary rewards.
9. Reward the winners by asking them to help the losers.
10. Pay attention to the details.
11. Experiment with the details.
12. Monitor many more measures than you use to motivate.
13. Try to create long-term support.

As with Behn's list of problems, my experiences in education lead me to validate this list of suggested practices. I believe that any of these suggestions would fit within the scheme of school operations.

I offer one additional suggestion for the list that I draw from Elmore's work. I see a powerful relationship of what Elmore terms "reciprocal accountability for capacity and performance" with real opportunities for making schools accountable and motivating educators through a well-defined measurement system. I suggest adding Elmore's reciprocal accountability to Behn's list of recommended practices.

I also offer one observation on Behn's comments concerning the practice of micromanagement in government. Behn suggests that in most public accountability systems, when an operational unit falls short of a performance target, policy makers typically react by increasing controls over that unit. The inverse, in Behn's view, is true when goals are met—controls typically are loosened where an operational unit goes beyond a performance target. In the case of education, Behn has hit the bull's eye. He has underscored one of the most significant reasons why educators typically regard top-down accountability measures with great disdain. From their perspective, unreasonable controls, which may result from making the complex simple, confusing data with information, or ignoring information in the heat of political battle, become impediments to achieving a well-intentioned goal. The result of this syndrome of increasing control and falling short may represent the great Catch-22 of the accountability systems they encounter.

ACCOUNTABILITY IN LOCAL SCHOOL DISTRICTS: LEARNING TO DO THE RIGHT THINGS

Elmore's paper suggests answers to the dilemma to which I confessed in my introduction. It is his view that most boards of education and superintendents do not formulate policy or allocate resources in ways that contribute to the ability of teachers and principals to improve teaching and learning. He suggests that their failure to do so is one of the reasons why state level policy makers have focused on schools, rather than school districts, as the units of operation upon which to target improvement efforts. The good news for boards and superintendents, according to Elmore, is that there is an important role for them to play, *if* they are willing to rethink their positions in the interplay of school organization and management.

Although reactors to papers presented in forums such as the O'Leary Symposium are often expected to challenge the logic, the research foundations, or the conclusions of the presenters, I strongly support Elmore's work. And, although superintendents are the focus of much of the criticism contained in Elmore's paper, I find it hard to disagree with his conclusions, especially when they offer me a meaningful opportunity to deal with my own professional dilemma.

During the past decade, I have engineered a great deal of planning time spent in the Springfield Public Schools on defining standards and on seeking a management structure that would empower teachers to be part of a process for improving student performance. Most of the focus on standards, learning goals, and outcomes has occurred in reaction to various state mandates or, in the very recent past, to the mandates of the local school board. The reactions of most teachers and principals to the need to define and re-define standards, goals, and learning targets have ranged from tepid enthusiasm to outright refusal to "do this stuff" again. Elmore most effectively describes the reluctance of teachers to embrace accountability efforts in the words of the kindergarten teacher who challenged the guys in the suits to get down on the floor with her children and see what is really going on in her classroom.

Parallel to our work on standards in the Springfield schools, we have moved toward site based decision making. Embodied in that movement is a search for the "silver bullet" of governance structures. When I read the following sentence in Elmore's paper, I had the sensation of looking into a mirror and seeing a frightening image that I had not expected to see: "Simply setting student performance standards, engaging in bureaucratic box-shuffling, and fussing with governance arrangements probably won't produce discernible effects on what teachers teach and what students learn." Bingo! Zap! Hello! If Elmore's work had ended there, I would have been in a position to see how I and the board of education have not been linked to the process of improving teaching and student performance. Had his paper ended there, however, I might have concluded that the link is not possible. The good news is that Elmore went on to describe what good schools do by "adding value to student performance in schools."

Elmore suggests that when schools *add value*, they increase their capacity to do the right things. What Elmore offers practitioners is a way to connect the issues of standards and governance to the improvement of teaching and learning. A source of past frustration and a future source of even greater frustration, as he points out, is the tendency of policy makers to believe that setting higher standards alone will lead to improvement. Adding to that frustration is the notion that empowerment of teachers lends a kind of magic to the context of governance structures, even if they are only empowered to keep on doing the same things that have *not* led to improvement. This is where Elmore lays out the case for knowing how and why to increase the capacity of schools to improve teaching and learning—for knowing how to do the right thing. By rethinking their roles and addressing the issue of increasing capacity, according to Elmore, boards of education and school administrators can become linked to improving teaching and learning.

As I consider Elmore's suggestions, I can react to both the reasons and the means for following them. I have observed in the past two decades a growing concern among teachers and principals about diminished student learning or failed attempts to improve it. During my entire career in education, I have worked in schools where professional staff have wanted to do the right thing. I have helped design and allocate resources in support of staff development efforts with very high expectations. Some have succeeded; others have operated in a less positive way, and a tremendous number have simply run time off the clock and drained district resources without result.

One of the successful efforts began four years ago when our district entered into a collaborative partnership with the Ball Foundation in the hope of creating a school that could accomplish the improvement of teaching and learning. The resources provided to our district and the collaborating school by the Ball Foundation and its partner, the University of Illinois' College of Education at Urbana-Champaign, have helped enhance our understanding of what good schools do. The quality, frequency, and focus of staff development at the collaborative school have significantly improved. Much of that inservice training has focused on governance, structure, and team building. The effort has resulted in the collaborative school's growth and improvement, but it would be fair to say that all of us associated with the effort are aware that we have not yet, in the words of G. Carl Ball, "hit the home run" we are looking for. We are close, but we are not yet there.

During the same period of time, the faculty of a second Springfield school embarked on an effort to redefine teaching and learning in their building. An approach that began with a team of four teachers and the support of the building principal and one central office curriculum specialist was grounded in a commitment to teaching teachers to teach differently. The commitment of this group to inservice training, first for themselves and later for their colleagues, has resulted in a revolution in teaching and learning that now affects the lives of students and teachers in nine district schools. Staff development and a focus on doing the right thing have been the primary foundations of these efforts; standards and governance issues have played secondary roles. Elmore's paper offers us a new view of the reasons for our success.

Another example I offer from recent experience to support the premises of Elmore's paper relates to the case he makes for staff development and training as the anchor of school administration. Elmore suggests that if local school systems are to promote teaching and learning in tangible ways, then central administrative offices will have to look and act very differently than most presently do. Having grown up professionally in a hierarchical central office, I have often wrestled with the counter-productivity that I've noticed in its compartmentalization. This compartmentalization of expertise and specialization has been at the center of my dilemma about linking the functions of the central office to meaningful school improvement. My long-term observations suggest that compartmentalization has produced roadblocks to communication and collaboration among central office staff members who compete with one another and who withhold information from one another and from personnel in the schools, the people who have the greatest need for it. The result has been a tremendous waste of time and resources and a loss of opportunities for school improvement.

With the concurrence of our board of education, we have recently restructured central office responsibilities in the Springfield district. Our goals in doing so include: (a) changing the culture of communications and decision making so that the needs of schools (students, teachers, and principals) define the work of central office staff; (b) placing a renewed emphasis on the mission of central office staff to provide support for and service to teachers and principals; (c) redesigning information systems within the district so that the needs of school level personnel drive the primary functions of that system; (d) beginning a reallocation of existing staff development resources to better align them with the work of school improvement teams; and (e) beginning a commitment to drive additional existing resources to the support of building level staff development initiatives. When restructuring is complete, the number of existing central office positions will be cut by approximately 25 percent, and the job responsibilities of central office staff will be altered in significant ways. The initiative to make these changes was already under way when Elmore's paper became a resource to us, but his position regarding the key function of staff development in the work of central office staff became an important element of our plan. The obvious intended outcome of this restructuring is to enhance the ability of the school system to assist in building the capacities of district schools to improve teaching and learning.

A final section of Elmore's work deserves mention. In his concluding pages, he proposed four principles for design and action: (a) the roots of accountability: responsibility and expectations, (b) existence proofs and the theory of small ties, (c) reciprocal accountability for capacity and performance, and (d) context matters. Each is vital to the successful transformation, but none is more critical than the principal of reciprocal accountability for capacity and performance. I have discussed this principle with a number of teachers, principals, and central office staff members. I plan to share it with the board of education. The notion that policy makers and administrators be held accountable to teachers and principals for providing a means to increase the schools' capacity to do the right thing commensurate with increased standards imposed on them is revolutionary to the practices of most school districts, including ours.

Without question, teachers and principals believe the principle of reciprocal accountability is absolutely essential to any approach to implementing an accountability system that works—one that produces real improvement in teaching and learning and one that delivers more than promises and platitudes. As Elmore predicts, the reaction of central office staff is mouth-gaping shock. Most have asked if I really think we can do such a thing. The answer to that question will determine whether or not the restructuring we have initiated makes any significant difference, or if it is just another exercise in futility.

CONCLUSION

Taken collectively, the papers presented by Behn and Elmore represent important views on the issues of school accountability and the financial management of the resources entrusted to school systems by their owners. Each paper underscores the belief that key participants in public education have important roles to play in the processes of accountability. Each paper clearly supports the contention that setting and raising standards alone will not produce the accountability being sought by various policy makers and stakeholders. Each, in its own way, reinforces the point that how and what we measure in the performance of students will be linked to whatever efforts we make to improve our product: teaching and learning.

Participation in the O'Leary Symposium has increased my personal capacity for framing possible answers to the question of how those who work in a district's central office can effectively contribute to improved teaching and learning. It has provided me with a better understanding of how I and the board of education can contribute to teaching and learning in meaningful ways. I hope that my thoughts offered here, along with the works upon which they reflect, provide a similar opportunity to others.

DEMOCRATIC ACCOUNTABILITY AND CITIZEN OWNERSHIP:

POLITICAL AND SYSTEMIC APPROACHES TO IMPROVING EDUCATIONAL PERFORMANCE

James G. Ward

An enduring issue in American education reform over the past two decades has been one of accountability. Policy makers have been concerned that the education system be accountable to the public and that this accountability be measurable in some tangible way. The underlying concept is that the public needs to be convinced it is getting its money's worth from public expenditures on public schools. Behn (1996) makes a convincing argument that accountability, established through a framework of measurement of performance and motivation to perform better, is critical in gaining new resources for public education. He is entirely correct in making the point, although these are not necessarily his words, that accountability initiatives need to alter the entire culture of the school system. Elmore (1996) makes an important contribution in viewing accountability as going beyond management

Advances in Educational Administration, Volume 5, pages 93-104.

ISBN: 0-7623-0081-7

to include questions of governance and goals. Both, however, focus on performance of the education system as a management issue and see accountability as matter of better organizational performance.

ACCOUNTABILITY IN A DEMOCRATIC SYSTEM

I would argue that there are at least two missing elements in this discussion and that these elements need to be addressed if we are to improve the educational system and establish accountability and legitimacy of that system. The first point is that accountability in a democratic polity must begin with the people, not with professional or elite group goals and purposes. The second point is that a model of public bureaucracy that has as its core the notion of citizen as owner is central to improving the system.

There is little question that the legitimacy of governmental institutions in all levels of government have come under attack in recent decades. Whether it be the resounding criticism of public education in *A Nation at Risk* in 1983, the reinventing government movement of the early 1990s, or the relentless attacks of conservative politicians on government as being too large and overreaching, there are demands that governmental agencies and institutions be eliminated, scaled back, reformed, or radically improved. The heart of the argument is a call for reductions in government budgets and greater accountability of public bureaucracies.

Rubin (1996) writes that the economic, social, and political context of government has caused different emphases in public budgeting at different times. In earlier times, the focus of public budgeting has been on financial control, managerial improvements, planning or prioritization. Rubin, however, argues that for the 1990s the focus of public budgeting is on accountability. Public managers who concentrate their efforts on better financial control, fostering managerial improvements, and the like may be effective managers, but they may be missing what is important to the public. The undermining of the legitimacy of public institutions in the political arena has led to a response by leaders of public institutions to restore public support and confidence.

Rubin (1996) also reminds us that accountability is a complex and many faceted set of ideas and that accountability has many meanings including responding to higher authority, reporting to the public, holding elected officials responsible for outcomes, and direct citizen control. Within the context of public education, we are not even sure which of these will work to our purposes, once we have settled on the purposes.

As Behn (1996) and Elmore (1996) have pointed out in their insightful and provocative papers in this volume, which address issues of how we provide greater accountability in schools, public education is no exception. Behn has connected accountability with motivation to ask how we can encourage people to provide better results. Elmore has challenged us to focus on *value added* to the student as a way to improve educational performance. Although both Behn and Elmore have offered sound suggestions and provocative ideas, they have missed some critical points.

We must make the distinction that accountability means both *accountability for* and *accountability to* (Bardach & Lesser, 1996). Before we can consider accountability for what outcomes, we must ask accountability to whom. Accountability to whom is not a simple question to ask, but in a democracy it must ultimately mean accountability to the people. Accountability is in the public interest. The purposes of accountability are to maintain democratic control over public bureaucracies and to ensure that the will of the people is carried out. When considering important questions of public policy, we must keep central to our consideration how we foster democracy, how we focus on the needs of society, how we deal with collective decision making for the community, and how we encourage learning for civic competence (Kirlin, 1996a). Behn and Elmore ask *how* questions, but maybe we need to look at *what* questions and *why* questions. Accountability in public education is not so much a question of public school management as it is a governance question and a value question and, therefore, a policy question (Frederickson, 1997; Hodgkinson, 1991).

ACCOUNTABILITY AND THE PURPOSES OF EDUCATION

The public, in a general way, wants to hold schools accountable for educational performance, but what comprises educational performance in the mind of the public? I submit that if we only look at student academic outcome measures, we have a limited and probably politically naive view of public schools and their public constituency. Jack Hennings, former U.S. House of Representatives General Counsel for Education, recently reported on the results of his odyssey around the United States talking to school people, parents, public officials, and others about education (Jennings, 1996). According to Jennings, "The public is ambivalent about children doing well in school. Most parents say that schools should have higher standards, but many parents are also reluctant to have their children work hard to achieve those standards" (p. 16). Jennings also reports that Americans are suspicious of too much learning, and he cites polling data from the Public Agenda Foundation that indicates that 23 percent of American parents would be satisfied if their child had numerous friends, led a well-balanced social life, and sustained a C-average in school. This hardly sounds like a clamor for greater academic performance and higher standards for schools. Jennings concludes on this point by writing:

> We clearly have a problem here. On one hand, the country's leaders say that schools must produce higher academic achievement, and the general public seems to agree, but on the other hand the parents want "well rounded" children who do modestly well in school but who also have friends and participate in social and athletic events. Which will it be? The schools are caught in the middle—they are condemned by leaders for not raising academic achievement and by parents if they do not have a good basketball team (p. 16).

My own experience in five years as a board of education member in a public school district validates and confirms the dilemma for schools that Jennings identifies. The public's expectations of schools are not always clear, and there are conflicts within a community about expectations. Parents often say that their first priority is a high quality education for their children, but then they qualify that statement by arguing for programs in athletics, in extracurricular programs in music and drama, or for shorter school days in hot weather. They may oppose a longer school year, stronger graduation requirements, or more homework, while at the same time decrying the educational performance of other nations. This anomaly exists and is part of the reality of public education. It is a mistake to assume that all members of a community have a common understanding of what they think the public schools should be about. Public schools are multifaceted institutions serving multiple goals.

An aspect of this phenomenon was identified by Steinberg and his colleagues (1996) in their study of more than 20,000 teenagers and their parents in nine high schools over a ten year period. Steinberg found that a major impediment to greater academic achievement in high school is the disengagement of students in learning and the belief that talent is more important than effort for academic success. Particularly pertinent to the point, however, is that the study found:

> Not surprisingly, parental lack of interest is strongly associated with children's academic difficulties and low school achievement. In addition, parental disengagement is a very good predictor of many of the problem behaviors whose levels have reached alarming proportions in the past twenty-five years: alcohol and drug abuse, delinquency and violence, suicide, and sexual precocity (p. 187).

The rest of the Steinberg's findings are also pertinent. In studying what he calls "perfectly average or above-average students attending perfectly average or above-average public schools," Steinberg summarized his research as follows:

> An extremely high proportion of American high school students do not take school, or their studies, seriously.... American students' time out of school is seldom spent in activities that reinforce what they are learning in their classes. More typically, their time and energy is focused on activities that compete with, rather than complement, their studies.... The adolescent peer culture in contemporary America demeans academic success and scorns students who try to do well in school. Schools are fighting a losing battle against a peer culture that disparages success.... Perhaps most serious, American parents are just as disengaged from schooling as their children are (pp. 18-19).

These conclusions were partially confirmed in field work done as part of a research project in which I am engaged on democratic leadership and rural youth sponsored by the W. K. Kellogg Foundation. We found high school students with a sense of emptiness and "nothingness" about their high school experiences and with a lack of engagement in the academic aspects of high school. One of the reasons students felt disengaged from high school was that they felt that their voices were not heard by adults and that their views did not matter to anyone.

The most recent Phi Delta Kappa/Gallup Poll on education produces a similar contradiction (Elam, Rose & Gallup, 1996). Respondents were asked to rate the importance of possible goals of the public schools beyond providing a basic education. The wording is important because most attempts at ensuring accountability measure student performance beyond a basic education. For what should we be holding schools accountable? In the Phi Delta Kappa/Gallup survey, those who were polled rated the purposes of public schools that were offered by the pollsters as follows, with the percentages being those rating the purpose as "very important" or "quite important": "to prepare students to be responsible citizens," 98 percent; "to help people become economically self-sufficient," 96 percent; "to promote cultural unity among all Americans," 86 percent; "to improve social conditions," 86 percent; "to increase people's happiness and enrich their lives culturally and intellectually," 84 percent; and "to minimize current inequities in education for certain minority groups," 76 percent (p. 56). The responses did not seem to vary much at all whether the respondents were public school parents, private school parents, or had no children in school.

On the same poll, people were asked, "Which one of the following would you prefer of an oldest child—that the child get A grades or that he or she make average grades and be active in extracurricular activities?" National totals show that 28 percent said "A grades," 60 percent said "average grades and extracurricular activities," and 9 percent volunteered "both," while 3 percent did not answer (Elam, Rose & Gallup, 1996, p. 56). Public school parents and private school parents responded identically, and those with no children in school provided the least support for the "A grades" choice. Public school parents, private school parents, and the general public do not seem to support high academic performance over other possible purposes of schools. Yet, high academic performance is at the core of accountability efforts.

If students and parents are not engaged and invested in schooling, and if parents are more satisfied with well-rounded children with modest academic records, what does this say for performance based accountability systems and attempts to motivate schools toward greater performance?

Professional Versus Public Control of Schools

One of my personal and professional interests for many years has been the issue of professional control versus public control of public schools. One working hypothesis might be that education professionals and large segments of the public may see the purposes of public schooling very differently from one another and, therefore, may have very different perceptions of what accountability means. We may be too quick to assume that if schools are to be accountable that they are to be accountable for improved academic performance. There seems to be a dissonance between how some policy makers, business leaders and educators are looking at school accountability and how parents and the public view it. This, I submit, is a critical problem.

Maybe we need to recognize that the fundamental problem of accountability in schools is that we are quite unsure as a polity on what we believe to be the purposes of schooling and that we need to discuss and debate that topic after we agree that it is a problem. Recognition of the problem is the first step in exploring ways to solve the problem. I am not sure that at this point we accept the proposition that a lack of any consensus on the purpose of schooling is one of our fundamental problems.

Mathews (1996) points out that Americans generally want to support public schools, but they are ambivalent about them. The public believes that the schools are hampered by social problems not of the schools' making, and they believe that not much can be done to improve this situation. He suggests that not only do schools need to be improved, but that the relationship between the school and the community needs attention. He goes on to cite a *Washington Monthly* article by Katherine Boo that argues that education reform has not been successful in the United States because reformers have not been able to master the process of change in schools, because citizen participation has been actively discouraged, and because the reforms have assumed that the long-standing commitment to public schools still exists (p. 17).

The Myth of Manageability

This leads to what I call the myth of manageability. If we think we have consensus on a particular public policy problem when in reality we do not, then we find it frustrating that the simple solutions proposed and implemented to solve what we assume to be a simple problem do not work. Freeman (1996) illustrates this phenomenon in discussing the causes and possible solutions to the growing income inequality in the United States. Each person interested in the issue seems to have a favorite cause and a favorite cure. Each of the suggested causes, however, do not seem to adequately explain the income equalities, and none of the proposed solutions seem adequate to effectuate increased income equality. As a result, there is a sense that we are ineffective in addressing the issue.

Part of the difficulty in the myth of manageability is that we assume a problem is manageable because we oversimplify the nature of the problem. When we realize how complex the problem is, there is a tendency to argue that we have proposed incorrect solutions rather than to place the blame on incorrect assumptions about the problem. Accountability in public education may seem elusive because we have made incorrect assumptions about that the public expects from public schools.

CITIZEN OWNERSHIP AND DISCOURSE

In order to address the issue of accountability in public schools, we may need to discard a management model and explore alternatives, such as a governance model. One avenue for exploration is the notion of the citizen as owner in a democratic polity. Schachter (1995) makes the contrast in our view of government between that of a customer oriented model and a model that emphasizes the citizen as owner. She argues that owners play an more active role in government, while customers are more reactive. Therefore, important functions of citizenship are education for citizenship, greater information exchange, and service learning for citizen development. Citizens should not be passive, as they often are in our education accountability debates. They should be active participants in articulating what they want and expect from schools and what schools need to do to establish stronger legitimacy.

> A customer-centered model puts citizens in a reactive role where they are limited to liking or disliking services and hoping the administrators will change delivery if enough customers object. Owners play a proactive role; they decide what the government agenda will be (Schachter, 1997, p. 9).

This is often difficult to achieve because in spite of a great deal of rhetoric about citizen participation, we often have low levels of citizen participation because political elites oppose it (Berry, Portney & Thomson, 1993). In public schools, educational professionals often work hard to dominate control of the public agenda and actively seek to deny the citizenry any meaningful influence over agenda setting. The model of citizen as consumer better fits a management model of government with control over institutions like public schools by professional elites who undermine attempts at increasing citizen participation in educational decision making. The model of citizen as owner is a threat to the cult of professionalism in schools.

In bolstering this argument, in an article based on her 1996 presidential address to the American Educational Research Association, Darling-Hammond writes:

> Repairing the torn social fabric that increasingly arrays one group against another will require creating an inclusive social dialogue in which individuals can converse form a public space that begins together diverse experiences and points of view. This suggests not only education *for* democracy, in the sense that we think of students needing to learn trades

> and good citizenship, but education *as* democracy—education that gives students access to social understanding developed by actually participating in a pluralistic community by talking and making decisions with one another and coming to understand multiple perspectives (p. 6, emphasis in the original).

Darling-Hammond goes on to suggest that most schools are not very good places in which to learn democracy. She notes also that we need to include parents and the community in this public discourse.

Kirlin (1996b), in a commentary on an earlier paper by Behn (1995), argues that an idea central to American public administration is:

> What are the instruments of collective action that remain responsible both to democratic political processes and to core societal values? ... At the policy level, a recurring choice is between instruments which rely more on elite judgment and authoritatively directed compliance versus those that emphasize citizen judgment and empowering strategies (p. 421).

Relying on both Darling-Hammond and Kirlin, we need to develop mechanisms whereby both the citizens and our students can create public space in which to engage in discourse about the purpose of public education and how public educational institutions can be accountable for achieving those purposes. Kirlin believes that Behn's question of how we can measure performance in public institutions can be addressed by asking, "How can processes of social learning be improved, including knowledge of choices available, of consequences of alternatives, and of how to achieve desired goals, most importantly, the nurturing and development of a democratic polity?" (p. 421). Kirlin is not disagreeing with Behn, but simply expanding on Behn's question.

Public Space for Public Discourse

An example of an attempt to create public space for public discourse around critical educational issues is the program of Public Agenda and the Institute for Educational Leadership to promote public discussion of important issues. Its first attempt in St. Paul, Minnesota, was reported recently in *Education Week* (Bradley, 1996). The purpose of the project, according to the news report, is to provide a process to assist communities to find solutions to educational problems while allowing the broadest possible diversity of voices to be heard. Citizens need to be able to meet as political equals in a forum that promotes open and free discussion, with the opportunity to have information as a free resource and to consider issues of importance to the community.

This supports what Mathews (1994, 1996) discusses in calling for people to increase civic involvement and build public communities in which they engage in dialogue to help solve public problems. Greater citizen involvement in important public policy matters not only provides guidance for public officials, but it builds support among the citizenry for public institutions. Important aspects of citizen involvement are "public forums and public discourse" about public policy issues (Mathews, 1994, p. 198). He contrasts the concept of "engaged citizenry" with that of a "persuaded population" and argues that we can sell ideas to people who may buy them, but this is inferior to developing a sense of ownership among people. With ownership, education becomes a community strategy rather than a school strategy (Mathews, 1996).

Discursive Democracy

Much has been written about deliberative or discursive democracy (see e.g., Dryzek, 1990; Fox & Miller, 1995). Such a deliberative or discursive model may provide a model to provide greater accountability through greater public participation in the processes whereby we determine for what schools should be held accountable and to whom. The function of government is not only service delivery, but also creation of effective governments to assist citizens in making collective choices and to make political choices defining and protecting the character of the community (Kirlin, 1996a). Citizens need assistance through some sort of a deliberative or discursive process to make collective choices about what is important, and education is clearly a key governmental function in defining and protecting the character of the community.

A discursive or deliberative approach emphasizes citizens' retention of their power of judgment over public matters with decisions being made by deliberative means rather than through the more traditional mechanisms of representative government and majoritarian rule, with limitations on the unchallenged authority and power of formal institutions and professional experts (Warren, 1996). Discursive or deliberative democracy requires an open forum for public discussion of important issues, where the participants are communicatively competent and participation is non-hierarchical and where unequal power situations are not allowed to exist in participation (Dryzek, 1990). Everyone listens, and everyone can speak. This approach to making public decisions is diametrically opposed to the usual mode of decision making in schools, where professional elites dominate in terms of active participation and access to expertise, where the system is hierarchical and dissent is repressed, and where unequal power situations are not only tolerated, but are imposed by the school professionals.

Information for Decision Making

Broom (1995) has found in her research on performance based programs in state government that a critical factor is the use of performance based information for budgeting. This requires altering budgeting information so that is more easily accessible to and readable by policy makers. Rubin (1996) discusses in more detail how governmental budgeting and financial reporting practices and documents need to change to provide greater citizen access to information for decision making. She writes:

> A budget that is oriented to accountability not only has more and different kinds of information in it, it is also closely related to actual events, it is not just a policy statement or wish list or a work plan without much concern for implementation. If the budget that is worked out in public view and that is widely circulated for approval is not the budget that is actually implemented, then there is not much accountability (p. 122).

Rubin presents a long list of strategies to get the public more involved in governmental decision making in order to promote greater accountability to the citizen owners of the government. She also notes how this may be inconvenient for the government, but is necessary for public accountability.

In the private sector making information more easily accessible has been recognized as a way to boost worker performance through helping workers understand the *why* as well as the *how* of changes in the workplace through a technique known as open-book management (Case, 1996). Open book management is a system and an attitude of sharing information broadly by making sure that every employee has all relevant financial information about an enterprise and are taught to understand it. It seems appropriate in the public sector where the citizen is owner.

Mechanisms for citizen decision making through a deliberative or discursive model depends on more than just information. Research in this area has found that improved accountability can be achieved through improved decision making and through the use of processes like citizen surveys, citizen panels, involving surveys, interviews, focus groups, and neighborhood meetings (Fox & Miller, 1995).

The model of citizen as owner in a discursive or deliberative democracy may be a promising approach for accountability for improved educational performance. Including the public in decision making is consistent with the basic tenets of a democratic polity. This is an area that is antithetical to the thinking of many school administrators, but it may well provide the key to meaningful reform.

CONCLUSION

The answer to the question of how we can improve educational performance through accountability may lie not so much with the accountability mechanisms we devise, but with our commitment to embrace the requirements of democracy and our willingness to engage in civil and public discourse in public space about the purposes of education and what we really expect from schools. To assume that we know or agree upon the purposes of education may not only lead us down false paths, but it may dash our expectations by invoking the myth of manageability and distorting our democratic processes.

In writing about budgeting for accountability, Rubin (1996) makes the point that "Government has an obligation to educate, to rouse, to interest, to involve the citizens if a sense of citizen ownership and pride in government is to be restored" (p. 127). Accountability starts with a sense of ownership. It involves citizen participation in decision making in a serious way. One of the elements of this model is that as public bureaucrats, we may need to listen to the public to hear what they have to say (Fox & Miller, 1995).

We need to discard the dysfunctional metaphor of the citizen as consumer, with its emphasis on structural reform and professional control of schools. In its place, we need to embrace the metaphor of citizen as owner and a view of schools that allows for democratic governance and control.

REFERENCES

Bardach, E., & Lesser, C. (1996). Accountability in human services collaboratives—for what? and to whom? *Journal of Public Administration Research and Theory, 6*, 197-224.

Berry, J. M., Portney, K. E., & Thomson, K. (1993). *The rebirth of urban democracy.* Washington, DC: Brookings Institute.

Behn, R. D. (1995). The big questions of public management. *Public Administration Review, 55*, 313-324.

Behn, R. D. (1996). *Linking measurement and motivation: A challenge for education.* Paper presented at the First Edwin J. O'Leary Symposium on School Financial Management, University of Illinois at Urbana-Champaign, October.

Bradley, A. (1996, October 2). Project seeks meetings of the mind on reform. *Education Week*, p. 3.

Broom, C. A. (1995). Performance-based government models: Building a track record. *Public Budgeting and Finance, 15* (4), 3-17.

Case, J. (1996). Opening the books. *Harvard Business Review*, 118-127.

Darling-Hammond, L. (1996). The right to learn and the advancement of teaching: Research, policy, and practice for democratic education. *Educational Researcher, 25* (6), 5-17.

Dryzek, J. S. (1990). *Discursive democracy: Politics, policy, and political science.* New York: Cambridge.

Elam, S. M., Rose, L. C., & Gallup, A. M. (1996). The 28th annual Phi Delta Kappa/Gallup Poll of the public's attitudes toward the public schools. *Phi Delta Kappan, 78* (1), 41-59.

Elmore, R. F. (1996). *Accountability in local school districts: Learning to do the right things.* Paper presented at the First Edwin J. O'Leary Symposium on School Financial Management, University of Illinois at Urbana-Champaign, October.

Fox, C. J., & Miller, H. T. (1995). *Postmodern public administration: Toward discourse.* Thousand Oaks, CA: Sage.

Frederickson, H. G. (1997). *The spirit of public administration.* San Francisco: Jossey-Bass.

Freeman, R. B. (1996). Toward an apartheid economy? *Harvard Business Review*, 114-121.

Hodgkinson, C. (1991). *Educational leadership: The moral art.* Albany, NY: State University of New York Press.

Jennings, J. F. (1996). Travels without Charley. *Phi Delta Kappan, 78* (1), 11-16.

Kirlin, J. J. (1996a). What government must do well: Creating value for society. *Journal of Public Administration Research and Theory, 6*, 161-185.

Kirlin, J. J. (1996b). The big questions of public administration in a democracy. *Public Administration Review, 56*, 416-423.

Mathews, D. (1994). *Politics for people: Finding a responsible public voice.* Urbana: University of Illinois Press.

Mathews, D. (1996). *Is there a public for public schools?* Dayton, OH: Kettering Foundation Press.
Rubin, I. (1996). Budgeting for accountability: Municipal budgeting for the 1990s. *Public Budgeting and Finance*, *16* (2), 112-132.
Schachter, H. L. (1995). Reinventing government or reinventing ourselves: Two models for improving government performance. *Public Administration Review*, *55*, 530-537.
Schachter, H. L. (1997). *Reinventing government or reinventing ourselves: The role of citizen owners in making a better government*. Albany, NY: State University of New York Press.
Steinberg, L. (1996). *Beyond the classroom*. New York: Simon & Schuster.
Warren, M. E. (1996). Deliberative democracy and authority. *American Political Science Review*, *90*, 46-60.

PART II

STATE-DRIVEN SYSTEMIC REFORM: CURRICULUM CONTENT STANDARDS

THE POLITICS OF DEVELOPING AND SUSTAINING MATHEMATICS AND SCIENCE CURRICULUM CONTENT STANDARDS

Michael W. Kirst and Robin L. Bird

DECIDING NATIONAL STANDARDS

As the national standards debate demonstrates, curriculum policy making is essentially a political as well as a technical process (Ravitch, 1995). Politics is the authoritative allocation of competing values. Curriculum disputes over such topics as AIDS education and creation science highlight the embedded values conflict (Wirt & Kirst, 1992). Proposals to increase curricular scope have reached their current conclusion when elementary teachers are expected to teach reading, writing, several varieties of arithmetic, geography, spelling, science, economics, music, art, foreign languages and history at the same time as they help children to develop physically, morally and intellectually, and mold them into good citizens.

Advances in Educational Administration, Volume 5, pages 107-132.

ISBN: 0-7623-0081-7

The oldest and simplest solution to deciding curriculum standards is to endow an individual or small group with the authority to make these decisions by exercising professional and presumably expert judgment (Massel & Kirst, 1994). This decision making body (e.g., a state school board or a national subject matter association) is linked to the community that gives it authority and power. This linkage provides the decision makers with a degree of autonomy that ranges from absolute responsiveness to virtual independence. For example, the Clinton administration's 1994 Goals 2000 law advocated that disputes concerning national standards content should be resolved by a new agency of the federal government. In 1997, President Clinton proposed a national test to be taken by all students in fourth grade reading and eighth grade mathematics.

But this proposal only pushes our search one step further. What sort of decision procedures do curricular decision makers follow? In the past, the procedure could best be described by what Lindblom and Braybrooke (1963) called a strategy of *disjointed incrementalism*, that is, a strategy that simplifies complex problems. The six major features of disjointed incrementalism are: (a) acceptance of the broad outlines of the existing situation with only marginal changes contemplated, (b) consideration of only a restricted variety of policy alternatives, excluding those entailing radical change, (c) consideration of only a restricted number of consequences for any given policy, (d) adjustment of objectives to politics as well as vice versa, (e) willingness to reformulate the problem as data become available and (f) serial analysis and piecemeal alteration rather than a single comprehensive overhaul. Many curricular decision makers use pragmatic methods of decision making that result in minimal changes at the margin. Conflict can be avoided by using vague language concerning standards and by covering so many topics that no major interest group feels left out. Content priority is sacrificed to the political necessity of coverage.

The push for national mathematics and science standards, however, proposes to change this disjointed incrementalism to a non-incremental reconceptualization and complete overhaul of subject matter standards and exams. Examples cited by national standards advocates include recent efforts by the National Council of Teachers of Mathematics and the National Academy of Sciences (Massell & Kirst, 1994). The politics of these efforts are complex, as one observer of the NCTM noted:

> Twin needs propelled the development of NCTM's standards for school mathematics: the need to gain consensus and the need to promote change. On the one hand, if these standards were to stand as the banners of the community, then they had to reflect shared values and commitment. On the other hand, if change was desired, then these standards had to do more than reflect current practice. New ideas were needed, ideas that departed from extant assumptions and practices (Ball, 1992, p. 9)

National, state, or local content standards require complex trade-offs, and there is no way to avoid conflict and a sense of winners and losers. Difficult choices must be made concerning content standards and the procedures by which these standards are set. Merely following the "right" procedural steps will not be sufficient. For instance, one constraint is school time, because not everything that all content advocates want can be taught in 180 days from 8:00 a.m. to 2:30 p.m. The history of curriculum politics has been one of jockeying for priority in an overcrowded school schedule. Some curricular priorities are politically organized into the curriculum while others are neglected. Historically, organized interests such as driver education and vocational education have been more effective politically than music education (Wirt & Kirst, 1975).

DETERMINING WHAT TO TEACH

Efforts to formulate curriculum standards have provoked conflict over the proper bases for deciding what to teach. Authoritative allocation of resources in these value conflicts is the essence of what we mean by the political process in curricular decision making. Should schools teach those things that are likely to be immediately useful in life outside the school or those things most fundamental to an understanding of organized knowledge? Should they emphasize the development of individuality or the transmission of a common cultural heritage? So long as people disagree on how to evaluate curriculum, they are bound to quarrel over its composition. The bases for this disagreement can be such things as social class or race.

From 1900 to 1970, four broad bases for evaluating curricular elements emerged as salient: tradition, science, community, and individual judgment. These values cohere people's preferences, but their conflicting nature creates political stress and demands for curricular changes. They are neither mutually exclusive nor exhaustive, but they do represent major streams of thought and feeling among curricular constituencies. In short, they are ways of answering Herbert Spencer's question, "What knowledge is of most worth?"

The *appeal to tradition*, exemplified in recent times by the Great Books program and the Council for Basic Education, rests on the assumption that subjects of study that survive the test of time are in the long view most beneficial and, therefore, should receive the highest priority in the curriculum. The *appeal to science*, the newest basis for curricular decision making, has received strong support from many influential groups, including the U.S. Department of Education. This appeal rests on the assumption that educational and psychological research will reveal cognitive concepts that should guide teaching.

The *appeal to community* presupposes that every school is part of a community of association and interests, in which reside the ultimate criteria of usefulness, relevance, and benefit of any curricular element. Therefore, those matters that deserve first priority in the curriculum are to be determined by the community, either directly, via its representatives, or indirectly by studies of the community. The *appeal to individual judgment* amounts to a skeptical denial of any rational value basis for curriculum making beyond the student's own values, needs, and desires as these are manifested in his own considered judgments. Adherents of this position argue that any basis for curriculum is doomed to failure if it purports to provide answers to Spencer's question.

Each of these values has its supporters and detractors who bolster their positions with techniques that we regard as political. Some schools stand primarily, and reasonably consistently, on only one of these bases. The curriculum of St. Johns University is based largely on the appeal to tradition, as are the curricula of a number of private "Latin" schools. A number of Christian groups advocate a traditional curriculum. By contrast, several schools embrace a scientific basis, such as constructivist pedagogy or stimulus and response psychological rewards. "Free" schools and "free" universities base their programs on free choices of individual students. Afro-centric schools are oriented to a community focus. By and large, however, U.S. public school programs are a heterogeneous mixture of these different bases. As such, they reflect the political compromises and diverse values found in any state or local district.

RECENT HISTORY OF POLITICS OF SCIENCE AND MATH REFORM

The political disputes of recent years were foreshadowed by post World War II developments. Until the 1950s math and science curricula were selected by individual school systems in response to perceived desires of local communities. The successful launch of Sputnik I in 1957 created demands for stronger federal and state roles using two broad strategies: more math and science at all levels and different content and instructional foci (Yee & Kirst, 1994).

Nevertheless, there were strong demands for preserving local control on some traditional curriculum matters and political conflict surrounding curriculum escalated in the 1970s (Dow, 1991). Efforts to ban books doubled in the first five years of the 1970s over the last five of the 1960s. A 1974 book banning crusade exploded into life threatening violence in Kanawha County, West Virginia. The American Library Association reported three to five episodes a week in 1981 ranging from the Idaho Falls banning of *One Flew Over the Cuckoo's Nest* to Anaheim, California, excising Richard Wright's *Black Boy*. After the Warsaw, Indiana, school board banned forty copies of *Values*

Clarification, the school board president posed the essential political question, "Who shall control the minds of the students?" (Wirt & Kirst, 1992).

One federal attempt to take a role in influencing local curriculum was rebuffed in the 1970s. Congress cut back the federal curriculum development role in large part because of a 1975 debate on the National Science Foundation's (NSF) proposed social curriculum, "Man: A Course of Study" (MACOS). Typical of this pressure was the charge of John Conlan (R. Arizona) that MACOS was a federal attempt to use classrooms for conditioning, to mold a new generation of Americans toward a repudiation of traditional values, behavior, and patriotic beliefs (Wirt & Kirst, 1992). Twenty years earlier, however, the federal government had entered the curriculum and text development field because critics alleged that schoolbooks were outdated, inaccurate, dull, and lacking in diversity. Scholars and experts on whom federal and state governments rely have been criticized for trying to impose their own cosmopolitan and secular values upon diverse local communities. Curricular reform itself has been professionalized through government and foundation grants. No longer are perceived crises such as Sputnik required to generate curriculum change, because curricular change now has a self-starting capacity.

Curricular conflict has many roots. Military threats or changes in public sentiment, such as the women's movement, generate value conflict about curriculum, even though this is not why these events were initiated. Other forces, such as court decisions favoring bilingual education or the pronouncements of influential individuals, can change the curriculum orientation without the direct development of new materials. In order to incorporate all these influences, the process of new textbook creation is "managed," whereby a writing team prepares a series of texts. The actual author is frequently the publisher's internal editor, not the authors listed on the title page. Teachers are also contributors to textbook content through their instructional preferences.

These skeletal concepts can be brought to life by the flesh of reality provided in a specific education issue. The reform of the mathematics curriculum in this country from the mid-1950s to the late 1970s nicely illustrates these multiple influences. It also shows the essentially political nature of curriculum decision making, even though professionals often regard curriculum as apolitical and something to be shielded from gross political concerns.

The traditional methods of teaching mathematics in public schools are rooted in early nineteenth century pedagogy. For some time, however, mathematicians and math teachers had been discontented with these methods. During this private issue stage of policy making, discontent was not sufficient to generate anything other than a scattered questioning of the old ways. Some university professors were experimenting with a new approach to math instruction, but there was as no sense of urgency or resource to act. The private-issue transformation stage of policy making occurred at the confluence of several dramatic events in the late 1960s.

Politics of the New Math: 1960-1970

Americans were enormously shocked in the late 1950s by the Russian success in orbiting the first space vehicle, the Sputnik. Critics of the American schools were already charging educators with poorly educating the modern generation; they seized on Sputnik as an illustration of how this American "defeat" was attributable to poor schools. Thus, external technological changes were used by reform groups who had been working for changes in math for many years (Dow, 1991). The media appealed to university curriculum specialists for enlightenment and parents made demands for action on state and national policy makers; the issue quickly became a public concern and was placed on the political agenda.

The crisis atmosphere of this issue's emergence illustrates how quickly a new national agenda can be created. Congress legislated and the U.S. Office of Education implemented a new law to provide funds for improving the quality of math and language training and for increasing the supply of such teachers. This National Defense Education Act (NDEA) of 1958 provided funds for developing new curricula in math, training employed teachers in the new math, and incorporating these changes into schools of education. In short, this policy output stage elevated the goal of providing students with better math training to national attention by redistributing federal money to teaching methods aimed at that goal.

The implementation stage was impressive in its thoroughness as Federal agencies fastened on a particular strategy of raising students' math comprehension. This strategy involved a new way of thinking about math theoretically, which could be taught from the first grade onward by methods that students could easily grasp. There was an attempt to blend the traditional curriculum ladder (arithmetic, geometry, beginning and advanced algebra, trigonometry and in some schools, calculus) in many localities. Students were introduced sequentially to different levels and relevant applications of such concepts as set theory. Thus, to get across the concept of numbering, instead of using the traditional method of counting to the base of ten, students were taught to count from the base of seven or another number. The logic of numbering was sought rather than the rote learning of a multiplication table.

Federal funds underwrote eight to ten week summer training institutes for many teachers. Special text materials were devised and disseminated; later, computer usages became available. States endorsed these innovations, providing additional funds to incorporate the new into the old math curriculum. Administrative state agencies beefed up their curriculum divisions relevant to math programs and assisted in the implementation of the changes. Local school boards were encouraged to participate by releasing their teachers for summer instruction and to graft the new onto the old. In time, if one's school system did not enjoy the new math, one felt left out of a national tide of change. Recall that such retraining and materials were provided for every level of schooling, and public and parochial schools alike were enlisted in moving the reform into the system and onto the students. This was all done in a remarkably few years.

The program outcome stage produced several results, but one of them was unintended and eventually counterproductive. One goal was to improve math training, and millions of students had been exposed to the new math concepts and understood them reasonably well. Another goal was to institutionalize the reform, and there were thousands of local education agencies (LEAs) that had adopted the new curricular concept. Yet another goal was to prepare students for the application of mathematics to life, and some feedback developed that was quite critical. Parents (who often did not themselves understand the new math concepts despite efforts to give them a quick exposure) increasingly complained that their children could not use mathematics for everyday requirements, such as multiplying or adding in preparing a bill or grocery list.

So widespread did this negative feedback become by the opening of the 1970s that state laws and regulations were altered. State text adoptions returned to the old math, and state tests changed accordingly. Signs that teachers were also reverting to the old math contributed to this reversal. There was little evidence that students were learning mathematical principles any better. With the widespread decline in math and English test scores across the nation in the mid-1970s, few curricula could claim much success, and the new math had no strong constituency within school systems or lobby organizations other than federal leadership through the NSF. It was not easy to monitor whether teachers stressed the new math, and the new math did not stimulate any structural identity within the school bureaucracy. By the mid-1970s, new national priorities for the disadvantaged and the political feedback cycle resulted in a shift away from the new math with little permanent residue. Some teachers continued to partially implement their NSF summer-institute in new math skills, but hostile external forces gradually diminished such content (Dow, 1991).

The Complexities of Systemic Change

Political developments in the 1960-1970 era underline the need to change the entire education system, but this strategy entails considerable political obstacles. The systemic reform strategy departs from the traditional practice of leaving content determination solely to individual teachers and local schools (Fuhrman, 1993). Frequent elections and turnover of key leaders make it very difficult for standards changes to persist long enough for systemwide implementation. The difficulty of re-electing federal and state leaders tends to preserve local control.

Historically, the educational system has not been designed for teaching a high level curriculum. Federal and state policies have tended to reinforce the use of textbooks and low level basic skills curricula that have become the de facto national curricula (Elmore & Fuhrman, 1994). For example, it is typical for schools and teachers to make their decisions about what they will teach by deferring to textbook publishers' tables of contents and to let standardized tests required by states and districts define the skills children should learn (Elmore & Fuhrman, 1994); students are required by the state to take statewide multiple choice tests and basic skills examinations that typically emphasize single correct answers and often do not adequately emphasize analysis, statistical inference, multi-step mathematical problem solving, hands on science, synthesis, expository writing and complex reading (Kirst, 1994). Neither texts nor tests encourage the development of high level skills (Archbald & Newman, 1988; Frederickson, 1984; Fuhrman, 1993; Tyson-Bernstein, 1988).

In addition, there are no high stakes for the students because teachers typically do not use students' scores on state assessments to determine grades. State tests are not used for college entrance, and employers rarely ask to see a high school transcript (Kirst, 1994). With the lack of explicit consensus about outcomes, low level skills that are familiar and relatively easy to teach become the de facto curriculum (Fuhrman, 1993; Smith & O'Day, 1991). Since the school system has no clear goals or directions, it cannot develop any authentic means by which to judge its progress, and thus, individual schools are left to use teacher grades.

Historically, policy makers have avoided making central determinations of outcome expectations because of support for local control. For the most part, educators have not tried to change the status quo, and many have argued against policy interference of this type in the past (Fuhrman, 1993). Educators have tended to be highly skeptical about policy makers' ability to develop ambitious and challenging standards for student achievement that can actually be carried out in the classroom. Their skepticism is based on the fact that it is easiest for politicians to set standards at levels that the current capacity can readily achieve. When standards are set too high, however, schools complain because they feel unfairly held to impossibly high standards (Ravitch, 1995).

Since many educators do not believe that the government is able to make enlightened education policy using top-down reform, they have placed their hope for educational improvement on individual school (bottom up) efforts (Cohen, 1983; Cuban, 1984; Elmore, 1990; Furhman, 1993; Purkey & Smith, 1983). As long as there is a school-based us-versus-them attitude at the policy level, systemic change may never be successfully achieved.

A third strategy integrates both top down and bottom up reform strategies, with both sides working together to help change the system. Project 2061 of the American Association for the Advancement of Science (AAAS) (1990), is a good example of the complex politics of systemic reform, as will be elaborated below.

PROJECT 2061: SCIENCE FOR ALL AMERICANS (SFAA): SYSTEMIC REFORM POLITICS

SFAA (AAAS, 1990) wrote, "Sweeping changes in the entire system from kindergarten through 12th grade will have to be made if the United States is to become a nation of scientifically literate citizens" (cited in Massell & Goertz, 1993, p. 3). *SFAA*'s primary goal is to make every American student science literate by the time he or she graduates from high school (Kirst, Anhalt, & Marine, 1997). Project 2061 completed Phase I in 1989 by forming literacy goals for science, mathematics, and technology; these are the understandings and habits of mind that the project felt are essential for all citizens in a scientifically literate society (AAAS, 1990). *SFAA* recommends that during students' elementary and secondary school career, their science education should focus on the following 10 themes, each of which includes several major sub topics:

The Nature of Science	The Human Organism
The Nature of Mathematics	Human Society
The Nature of Technology	The Designed World
The Physical Setting	The Mathematical World
The Living Environment	Historical Perspectives

By identifying these broad themes, Project 2061 differentiated itself from traditional science curricula. "The amount of detail that students are required to know, according to *SFAA,* is considerably less than in traditional science, mathematics, and technology courses. In order to provide a foundation for a lifetime of scientific understanding, ideas and thinking skills are emphasized in Project 2061 at the expense of specialized vocabulary and memorized procedures" (AAAS, 1990, p. 5).

In addition, although many of the topics recommended by *SFAA* are already taught in most schools, "*SFAA*'s recommendations differ from standard practice in their attempts to soften boundaries between traditional scientific and other academic disciplines, emphasizing connections instead" (AAAS, 1990, p. 6). First of all, *SFAA* suggests that through students' academic careers, when they study the ten themes, their science education should revolve around six major organizing concepts: systems, models, constancy, patterns of change, evolution, and scale; students should also understand how these concepts cut across disciplinary boundaries (p. 6). Second, *SFAA* aims to design a science framework that will lead students to develop three so-called "habits of mind": curiosity, openness to new ideas, and skepticism (p. 6).

Based on prior curriculum reform attempts, six controversial areas of *SFAA* appear to be:

(a) interdisciplinary connections that include science, math, technology, and social studies; (b) less content coverage and more in-depth analytical learning; (c) more science in elementary school; (d) greater emphasis on real world application; (e) more math, science, and technology for non- college bound students; and (f) increased use of heterogeneous groups and decreased use of tracking (Tyack & Cuban, 1995; Kirst, Anhalt, & Marine, 1997).

All of these components involve significant conflicts and resistance from organized groups.

Math groups do not want to be subsumed within the science curriculum, and vocational educators are trying to feature technology education as their own, not as a part of science. The pressures for content coverage rather than in-depth topics are inherent in demands by university admissions and AP exams. The elementary school day is so crowded that some content or activity must be eliminated if science time is increased. Many attempts to detrack schools have caused resistance from parents of higher achieving pupils (Kirst, Anhalt, & Marine, 1997).

"Project 2061 has been a reform effort that prizes flexibility and openness in its developmental process" and "as a result, it is a growing organism continuously reshaping itself in response to new information" (Massell & Goertz, 1993, p. 2). When curriculum is more flexible and open-ended, however, teachers need to become more active decision makers in the curriculum process than they currently are (Massell & Goertz, 1993). But teaching for understanding requires teachers to teach in a way that they were seldom taught (Elmore, 1993) and frequently goes against the grain of school organization variables such as 45 minute periods (Massell & Goertz, 1993, p. 30).

The Politics of Explicit Science and Math Standards

Today, math and science curriculum developers are bridging the gap between legislation and the classroom by specifying content and performance standards, while trying to give teachers and local school districts a meaningful zone of local discretion over how to achieve the goals of the legislation (Elmore & Fuhrman, 1994; O'Day & Smith, 1993; Smith & O'Day, 1991). There are four main areas of political tension that make it difficult to develop a supportive coalition inside and outside the schools for new math and science curriculum standards: (a) tension between leadership and political consensus; (b) tension between flexible and specific standards; (c) tension between up to date, dynamic standards and reasonable expectations for change in the system; and (d) tension between professional leadership and public understanding of what new standards will entail (Massell, 1994b).

Tension Between Leadership and Political Consensus

Previous educational reform efforts, especially large-scale curriculum reforms, have often been criticized for ignoring the social, political, and technical realities of implementation in schools and classrooms (Dow, 1991; McLaughlin, 1991; Yee & Kirst, 1994). The new mathematics and new science projects that were sponsored by the NSF from the 1950s to the 1970s are good examples of programs that were criticized because parents, teachers, community leaders, administrators, and others "had only limited, if any, involvement in the development of the new curriculum; were uninformed about the changes they were expected to make; and were ill-prepared to defend the reforms when challenges arose at the local levels" (Massell, 1994b, pp. 186-187). Because of the failed reform efforts of the past, educators today are well aware of the types of problems that will arise if notions of change are not widely shared at the community level (Carlson, 1995). Thus, most of the standards projects today try to gather inverse imput by engaging professional educators, community members, and others who have an interest in the standards. Gathering diverse input alone, however, will not achieve the development and implementation of leading-edge content standards, because leading edge content standards are frequently at odds with reaching broad consensus. Many groups have tried to develop consensus around content standards for their field, and even NCTM, one of the most successful, has been recently subjected to intense disputes.

Nevertheless, NCTM has been fortunate enough to achieve a high degree of initial consensus around the content standards that it designed (Massell, 1994b). The following are some of the processes that NCTM used to enhance the impact of the standards:

1. A long period of preparation prior to convening the standards-writing committees to lay some of the intellectual groundwork for mathematics reform.
2. Broad involvement in the development process. In contrast to past large-scale curriculum reforms, NCTM engaged more educators as well as subject matter specialists on its drafting committees.
3. A far reaching review and feedback. The organization embarked on an extensive consensus building process that involved thousands of practitioners, academics, and other professionals as well as members of the lay public in different stages of setting the agenda. NCTM received the endorsement of major professional associations prior to the release of its standards document.
4. An extensive consensus and capacity-building effort after the standards were drafted (Massell, 1994a, 1994b).

One should not be misled to think that science organizations will be successful at achieving content standards for their specific disciplines just by following the same steps that NCTM followed to formulate their standards (Collins, 1995). Although the aforementioned processes do seem to have been critical for NCTM to achieve consensus during content standard formulation, there were also other factors that played an important role in NCTM's success, many of which are unique to the discipline of mathematics (Massell, 1994b).

Massell (1994b) describes the uniqueness of mathematics that seemed to aide NCTM in the formulation of its standards in the following manner:

> Mathematics, unlike science, is not fragmented into a large number of competitive sub-disciplines; furthermore, the sub-areas that do exist (i.e., geometry, algebra, calculus) share a common conceptual base and language that facilitates discussions across them and makes goals like 'depth over breadth' more easily achieved. In contrast to science, mathematics does not tend to galvanize debate on pressing social issues or political concerns. The mathematics community has relatively few national organizations, and many have overlapping membership. These elements strengthen communication and provide a more solid foundation for consensus (p. 188).

The less turbulent political atmosphere in regard to standards at the time when NCTM developed its standards in 1988, compared to 1995, also facilitated its success. In fact, NCTM had to use its own resources to produce the standards because "federal and foundation actors did not think a national curriculum document was a good idea" (Massell, 1994b, p. 188). Since NCTM had no external support, it had to construct support for the standards project from within its membership. NCTM was also able to take its time (seven to nine years, depending on how you count) to develop its standards; a luxury today's standards developers do not have.

Although NCTM, by having many factors in its favor and by following the processes it did, was able to achieve a degree of consensus around the content standards it designed, one cannot ignore the fact that there were professional disputes that arose during the formulation process that still linger today. For example, NCTM members have paradigmatic arguments about how basic skills content should be taught, along with disagreements over when problem solving should be introduced (Massell, 1994a).

Members who took a behavioral approach argued for a teacher centered direct instructional method in which stressing memorization. Behaviorists also argued: You have to crawl/walk before you can run. If formulas aren't memorized, there will be no basis for the mathematical reasoning. If there is no mechanistic answer finding, there will be no conjecturing, inventing, and problem solving. If you don't know a body of so-called isolated concepts and procedures, there won't be any connecting mathematics and its applications. Judicious use of old-fashioned rote memory and drill are as necessary today as they were in generations past (Carlson, 1995).

Other NCTM members thought that classrooms should be student centered with emphasis on mathematical reasoning learned through constructing and solving problems. Their timing argument, and the one that NCTM accepts, is that skills and concepts can and do emerge during the process of problem solving and should proceed in tandem (Massell, 1994a). The *Curriculum and Evaluation Standards* states:

> Two general principles have guided our descriptions [of students activities related to math]: First, activities should grow out of problem situations; and second, learning occurs through active as well as passive involvement with mathematics. Traditional teaching emphases on practice in manipulating expressions and practicing algorithms as a precursor to solving problems ignore the fact that knowledge often emerges from the problems. This suggests that instead of the expectation that skill in computation should precede word problems, experience with problems helps develop the ability to compute (NCTM, 1989, p. 9).

Even if current science standards developers follow the processes that NCTM used and are aware of paradigmatic conflicts that may arise within their discipline, diverse fields like science did not try to develop standards in the same way as NCTM did (Collins, 1995). Current science standards developers were not afforded the time they need to build discussion and agreement among their diverse members. When NCTM developed their standards, there were no political or time pressures under which they had to work. Science standards projects, like most other standards projects today, do not have the luxury of taking seven to nine years to achieve initial consensus, but rather are forced to adhere to strict and short time lines. NCTM standards are under widespread political attack for their alleged emphasis upon using calculators and not providing enough training in logic and accuracy. Moreover, parents, employers, colleges, and vocational educators all have conflicting views about math content priorities.

Realizing the improbability of quick consensus among their diverse members and the fact that the high stakes market place of public ideas, professional status, and dollars may make dialogue and consensus a more difficult challenge, some of the current initiatives

> have set out to develop national standards in their field without widespread agreement among their members. The actions of these standards initiatives can provide the various fields with a wonderful opportunity for broad based discussions that may help to clarify the goals and purposes of the fields being discussed. With this in mind, future math and science standards developers may find it necessary to approve standards that not everyone will support in order to achieve the goal of producing leadership in the field; they may even find that they must compromise certain aspects of the standards they want in order to survive political challenges (Massell, 1994b, p. 189).

Tension Between Flexible and Specific Standards

Historically, the United States has moved from local control of curriculum to increased roles for state and federal governments. Thus, for national standards in math and science to be accepted, the people of this nation must want to have them, and the standards must be flexible enough to allow for local elaboration and variation. Keeping this in mind, Smith and O'Day (1991) propose simultaneously to:

> increase coherence in the system through centralized coordination and increase professional discretion at the school site. Thus while schools have the ultimate responsibility to educate thoughtful, competent, and responsible citizens, the state—representing the public—has the responsibility to define what 'thoughtful, competent, and responsible citizens' will mean in the coming decade and century (p. 254).

Today, the term *standard* is typically used as a sort of flag that reflects the valued goals around which educators can rally and decide for themselves how the objectives of the standards will be accomplished in their schools and classrooms (Ravitch, 1995). Thus, the new standards developers for math and science may find it undesirable to enforce a particular set of practices or materials. Rather, math and science standards could be designed to direct and guide local choice instead of determining and prescribing practice and teaching. No rigid or specific implications for practice would be inferred from standards (Ball, 1992; Myers, 1994; Sykes & Plastrik, 1992).

By being this flexible, math and science standards developers may run into the problem of not knowing when their standards have become so flexible that they can no longer provide leadership (Massell, 1994b). The same concern over flexibility emerged among experts who were trying to design teacher knowledge assessments for the National Board for Professional Teaching Standards:

> By not creating standards at what we would call a fine-grained level... standard writers leave the critical work of operationalizing standards for exercises and judging to the assessment developers. We, not the standards committee, ... imagined the vignettes or examples of accomplished teaching, we attempted to ground the standards in research, and we think the standards committee should have been involved in the assessment effort to operationalize standards (Pence & Petrosky, 1992, p. 12).

Therefore, "a certain level of detail in the content standard is necessary to guide the construction of performance standards, which will then guide test specifications, and finally the development of the tests themselves" (Massell, 1994b, p. 192). If the math and science content standards do not provide sufficient detail, then they will not pave the way for "other policy components such as assessment and instructional materials"; this is the same problem math and science frameworks are struggling with today (Massell, 1994b, p. 192). Thus, the national debate on math and science standards

> must go beyond generally worded standards to include the development of curricula specific enough to guide teaching and assessment. These must be the first steps; a syllabus-based examination system will have to wait until standards are established, because we cannot insure that students have a fair chance to learn what is tested until we have a curriculum in place (Koretz, Madaus, Haertel, & Beaton, 1992, p. 12).

But with more specificity comes less flexibility for people at various levels in the system and more political objections to policy makers who do not work in classrooms, telling teachers what to teach. Developing numerous, relatively detailed, and precise strands of content keyed to a common set of standards is a possible resolution to the flexibility/specificity dilemma.

Tension Between Up to Date, Dynamic Standards and Reasonable Expectations for Change in the System

In its 1992 report, "the National Committee on Educational Standards and Testing called for dynamic content standards, that is, standards that are continuously updated to reflect advances in scholarship" (Massell, 1994b, p. 193). But, how often should standards be revised? We already know how difficult it is to develop standards, and with all the interlocking systemic components like teacher repertoires, instructional materials, and assessments that take time and significant resources to develop, frequent revisions are not practical. California currently revises its curriculum frameworks in a subject area on a staggered, eight-year schedule. That is, one particular content area framework is reviewed every eight years, with a new subject being addressed each year by the state policy makers. Eight years may seem like a long time, but breaking the process up into the time it takes to complete each step demonstrates that eight years can become a very short time. For example, it takes approximately two years to revise the curriculum framework, and then publishers must be given enough time to align their texts accordingly. Furthermore, staff development, assessment, and other facets of the system must be constructed and implemented in the schools.

California has been frequently criticized for having its frameworks and assessments ready before the staff development and curriculum materials were in place (Massell, 1994b). Although the staggered schedule does not present a great burden of change on junior high, middle school, and high school teachers, it does for elementary teachers because they are responsible for a whole range of subjects and, therefore, have to reassess a key component of their curriculum every year (Massell, 1994b); district curriculum supervisors have the same problem (Marsh & Odden, 1991). Thus, math and science standards developers must keep in mind a practicality and feasibility time line; revised standards do nothing to help educate students if educational resources and systems cannot keep up with them.

Many educational frameworks, standards, and materials for mathematics, science, and other disciplines are also criticized because they tend to cover a wide range of material superficially instead of emphasizing in-depth learning of key concepts. Typically, such broad and vague frameworks do not push publishers to develop high quality material, have little impact in the classroom, and are "seen as little more than 'good doorstops'" (Massell, 1994a, p. 119). When they are provided to districts, schools, and teachers, they have little effect because of their intentionally vague language.

The politics that emerge between and among discipline members is a large factor that leads developers to focus on breadth rather than depth. For example, when frameworks and standards are being developed for a field, such as science that has various sub areas (e.g., biology, chemistry, physics), it is hard to get developers to reach a consensus concerning what should be covered in depth (Collins, 1995). Instead of working together and compromising so as to develop in-depth coverage of a few key topics, developers tend to include everything. By using broad and vague terms the developed sub areas (like physics or biology) do not offend, because topics from one area do not get emphasized more than another.

Tension Between Professional Leadership and Public Understanding of What New Standards will Entail

Parents, religious groups and other factions in society will continue to have an effect on the wording and contents of standards documents and on what teachers choose to teach in their classrooms. Challenges to public school programs and materials are nothing new in America. As McCarthy (1993) notes:

> For decades civil rights groups have been challenging materials as racist or sexist or as curtailing free speech; consumer groups have been contesting materials that promote bad health habits; environmentalists have been critical of texts that do not encourage global responsibility; and parents and religious groups have objected to the language and orientation of particular books, courses, and activities (p. 55).

Recent disputes, however, are different from the above mentioned challenges in many ways. In the 1960s and 1970s, "the most vociferous critics of public education usually came from the political left" (Whitty, 1984, p. 52). Since the 1980s, however, more and more critics of public education have come from conservative citizen groups; the groups have been labeled as the "new right." Also, the number of challenges has increased dramatically, and the central focus of the challenges has broadened from single books to entire programs and strategies to redesign schooling (McCarthy, 1995). In addition, the disputes have galvanized considerable media coverage. "Battle lines are clearly drawn, misinformation abounds, vicious accusations are being hurled, logic is often replaced by emotion, and there seems to be little desire to compromise" (McCarthy, 1995, p. 55).

The conflicts today seem to reflect the different values and world views of educators, parents, business leaders, and policymakers in regard to not only the purpose of schooling, but also the relationship between individual and societal interests in a democratic society (McCarthy, 1995; Marzano, 1993-1994).

GROUPS FAVORING TRADITIONAL CURRICULUM

Content standards based on outcomes of what students know and can do are confronting strong counterpressures. Outcome based education (OBE) in California, Pennsylvania, Iowa, Ohio, Virginia, and Oklahoma was amended or eliminated from statewide math and science content standards (Brockett, 1993). One of the nation's most outspoken OBE critics is Peg Luksik, who ran a close race in the 1992 Pennsylvania Republican gubernatorial primary. Luksik calls OBE a "sham" and says the basic philosophy that all students can learn "is just rhetoric that will guarantee a standard of lowest academic achievement." And, Ben Bancroft of Oklahoma Families Restoring Excellence in Education charges that "learner outcomes test values instead of academics, and emphasize self-esteem and problem solving, at the expense of knowledge in specific subjects" (Brockett, 1993, p. 8). The objection to problem solving is a significant problem for math and science content standards formulated under contracts made in 1992 by the U.S. Department of Education. All of these contractors stress problem solving.

Organizations that favor a limited federal or state government role in school curriculum, such as the Eagle Forum (Phyllis Schlafly) and conservative Christian churches (Pat Robertson), are being opposed by statewide big business groups such as the Business Roundtable. Business groups are concerned that local standards will not provide a well prepared workforce. The right-of-center movement, however, is growing in strength and forming an impressive national political network.

Some of the most active groups are the American Coalition for Traditional Values, the American Family Association, the Christian Coalition, Concerned Women for America, Eagle Forum, Educational Research Analysts, Family Research Council, Focus on the Family, and the National Association of Christian Educators/Citizens for Excellence in Education (NACE/CEE). Most of these groups are considered part of a loose coalition that emerged in the early 1980s.

CEE has a simple agenda: to return academic excellence to our schools, to return moral sanity and family values to our schools, and to elect parents to community and statewide school boards who will hire parent sensitive superintendents when these current boards don't listen. The primary targets of

CEE are OBE and liberal organizations such as the American Civil Liberties Union, People for the American Way, and the National Education Association (Ballen, 1995). According to its mission statement, CEE wants to shift the balance of influence in public schools to parents and local communities through grassroots programs and publicity and sees teaching about religion in public schools as an essential component to a complete education (Zahorchak & Boyd, 1994).

The agenda of Focus on the Family is very much like those of the Christian Coalition and NACE/CEE, but its primary reason for existence is to help preserve traditional values and the institution of the family. Focus on the Family supports high standards of success for all students as they learn objectives and are then evaluated by objective tests that do not rely upon subjective judgments of assessors (Ballen, 1995).

STRATEGIES OF TRADITIONAL CURRICULUM GROUPS

When trying to influence the content of public school curriculum, traditional curriculum supporters use a number of strategies ranging from litigation to personal persuasion (McCarthy, 1995). These groups have asserted that some materials promote either Christianity or an opposing creed and have even charged that certain books and instructional components (e.g., sex education, evolution) are anti-theistic. These groups have been unsuccessful in litigation where they have alleged that specific materials unconstitutionally advance an antitheistic creed, and courts have found the challenged books and instructional components to be religiously neutral (*Grove v. Mead School District*, 1985; *Mozert v. Hawkins County Public Schools*, 1987; McCarthy, 1993). Although courts uphold school boards that defend instructional programs, they also uphold boards that are restricting the curriculum, as long as the boards follow their own adopted procedures and base their decisions on legitimate pedagogical concerns, even if the board bases their decisions on traditional religious beliefs (*Hazelwood School District v. Kuhlmeier*, 1988; *Virgil v. School Board of Columbia County, Florida*, 1989; *Zykan v. Warsaw Community School Corporation*, 1980; McCarthy, 1993).

As these traditional groups know that there will be judicial deference to school boards, they have turned their central attention to influencing school board policies and elections. They successfully employ grass-roots strategies, such as door to door campaigns, to rally communities behind their attacks on particular aspects of public school programs (McCarthy, 1993). They effectively use democratic processes, "sometimes joining forces with taxpayers, senior citizens, and other conservative religious groups that share their agenda" (Arocha, 1993, p. 9). With such efforts, they have been very successful at strengthening their role in school board elections claiming to have elected close to 10,000 school board members nationally since 1989 (McCarthy, 1993).

Traditional groups have been successful at recruiting parents as part of their grass-roots strategies. A good example was the campaign against OBE orchestrated by Peg Luksik that included hundreds of Pennsylvania town meetings and a widely distributed videotape entitled *Who Controls Our Children* (McCarthy, 1993). The campaign was very successful at convincing a large number of people that OBE was part of a federal conspiracy associated with invading their children's privacy, "stressing values over academic content, and holding students accountable for goals that are so vague and fuzzy they can't be assessed at all" (Olson, 1993, p. 25).

In addition, many seemingly unrelated statewide school restructuring efforts (e.g., Florida's Blueprint 2000, Indiana 2000) "have been attacked under the OBE umbrella because they include standards and assessments that emphasize problem solving and independent thinking" (McCarthy, 1995, p. 12). These groups successfully recruit parents and other voters because many of their allegations "tap into... legitimate fears shared by many American[s]... who don't trust either government or schools to do what's right for... [America's] children" (Barth, 1993, p. 43). Critics of the government and schools "feel that current reform efforts are wrongheaded and [they] are fearful of radically different schools and the potential for technology to be used to invade students' privacy" (McCarthy, 1995, p. 12).

A key strategy for science and math standards developers is to talk with parents and to gain their trust. The average person does not take the time to read policies, legislation, or curriculum frameworks, but instead, relies on the interpretations of local concerned citizens, groups, and political officials for understanding documents. Historically, math and science standards developers have not taken sufficient time to talk to parents to help them understand the true nature of standards (Dow, 1991). Public engagement, however, cannot resolve deeply held value conflicts about curriculum content priorities. When various groups rally for or against standards change, developers "may decide that the educational benefits of the programs or topics are not worth the risk of an emotional community controversy" (McCarthy, 1995, p. 13). Developers and educators may even avoid topics that they feel might arouse controversy from the particular parent groups and factions. This risky strategy may merely delay political reaction.

SUMMARY OF CONFLICTS IN DETERMINING CURRICULUM STANDARDS

This paper has addressed numerous political issues concerning math and science standards. As a summary, we present several dilemmas and trade-offs policymakers and educators will have to face when trying to develop nonincremental content standards (Kirst, 1994). The list is written from the vantage point of the decisions that will be faced by any group when it deliberates about math and science content standards.

1. Who must be involved in the process to feel it is inclusive? Students? Business? If you exclude groups, you will be charged with bias. If you include every group that is suggested, you will have a cumbersome and slow process.
2. If you choose standards that achieve a broad consensus in the field, the "leading edge thinkers" will object. You will be accused of certifying "what is" rather than "what ought."
3. If you choose a standard that achieves consensus in a field you will not be able to satisfy demands for "less is more," because consensus expands topics rather than reducing them.
4. If you choose a standard that reflects current content consensus, you will be criticized for not sufficiently stressing interdisciplinary content. There is limited interdisciplinary content in standards, like NCTM recommends.
5. If you approve standards that are too general or do not contain pedagogy, you will be criticized for promising insufficient instructional guidance for teachers, and the content gaps will be filled by tests or assessments. If you do approve pedagogy or detailed standards, you will be criticized because standards are too long, too complex, and overly control local teaching practice.
6. If you do not hear appeals from the public for specific content changes (e.g., inclusion of creation science), then you will be criticized for not having pubic participation at the highest level and for leaving crucial decisions to a technical panel of non-elected officials. If all these protests are considered, the process will become bogged down in time consuming and fractious disputes.
7. If you approve similar structure or dimensions for all subjects, you will be criticized for ignoring the big (structural) differences between such fields as math, science, and social studies.
8. If you hear appeals from subject matter specialties (too much physics, not enough biology), you will end up arbitrating balance among technical fields within a subject matter area. If you do not hear these issues, you will be accused of approving unbalanced standards within a subject area. For example, there is too much physics and not enough biology in science content standards.

LOCAL AND STATE POLICY INFLUENTIALS IN STANDARD SETTING

Math and science political conflicts must be decided by some authoritative political body (e.g., state or local government entities), if they are to become

<u>*State Level*</u>

<u>*Insiders*</u>
Professional associations in science, math, and social studies, NSTA, AAAS, American Chemical Society, etc.
Higher education policymakers and professors in science and related disciplines.
State curriculum framework policy makers.
Textbook publishers/testing agencies (private industry).
NSF Curriculum Development.
National Education Goals Panel, including the proposed National Assessment and Improvement Council.
National Research Council.
Leaders in education policy in the legislature.
<u>*Near Circle*</u>
State Board of Education (Text Adoption States)
Teacher preparation institutions + teacher certification, National Council for Accreditation of Teacher Education, National Board for Professional Teaching Standards, and American Association for Teacher Education.
Ideological interest groups (e.g., creationists).
OERI in the U.S. Department of Education, NSF.
<u>*Far Circle*</u>
Organizations of state officials, National Governors' Association, Education Commission of the States.
Council of Chief State School Officers.
<u>*Sometimes Players*</u>
School accrediting agencies (e.g., North Central).
Business organizations, minority organizations.

Notes: Categories adapted from Marshall, Mitchell and Wirt, F. (1989). Rankings based on review of literature concerning the politics of curriculum policy making.

Figure 1.

official policy. The critical state decision makers for math and science standards will be different from those for policy in general and dependent on the political traditions within a particular state and local context. To understand the politics of policy making, it is necessary to highlight the key actors. For example, many insiders and crucial players for science standards are "sometimes players" in general state policy analysis. Our rankings for science curriculum influentials at the state level is presented in Figure 1. The ordering is from the most to least influential in terms of being able to help determine state and local science policies.

Figure 1 suggests science standards reformers should focus on the insiders and the near circle and devote less attention to the far circle and sometimes players. Business groups, however, can be mobilized behind science/math reform and are quite influential. There are also many crucial players at the local level that will help determine the success or failure of any standards proposal.

Key Players in Standard Setting at the Local Level	
Local School Boards	Parent Organizations
Teacher Unions	LEA Curriculum Supervisors
Administrators	Department Chairs

Some of the key local influentials are not organized in any way above the school or school district level (e.g., department chairs). They can be crucial in a specific locality, but they are very difficult to target in a general reform strategy.

FINAL THOUGHTS

Math and science content standards and exams are another chapter in the long running saga of U.S. curriculum politics. Decisions on what knowledge is most worth knowing are at the center of school politics, even though school finance usually attracts more media attention. Curricular standards are the crucial component of the overall vision of systemic reform (Smith & O'Day, 1991). Content standards are a beginning for subsequent state and local policy alignment of textbooks, assessment, staff development, categorical programs, and accreditation. All of these policy areas must be linked to teaching the national content standards in America's classrooms for systemic reform to succeed. Consequently, standards for math and science content are high-stakes politics.

Some potentially useful political strategies for math and science curriculum developers include better public engagement and parent involvement; coalition building with business, higher education, vocational education, and teacher organizations; and a recognition that some value conflicts are so deep that no reasonable compromise is possible.

In its 1993 publication, "Transforming Education: Overcoming Barriers," The National Governors' Association offered six guidelines for governors pursuing systemic education reform. These guidelines offer insight for educators and policy makers at all levels who want to pursue the development of standards:

1. *Send clear and consistent signals.* Encourage review and removal of all policies inconsistent with systemic reform. Keep new policies to a minimum and consistent with a clearly stated policy for transforming education. Be sure there are clear, two-way channels of communication throughout the system.
2. *Give priority to professional development.* Help redefine it as an ongoing part of the job of teaching, not just an occasional activity oworkshop.

3. *Balance top down and bottom up strategies.* Strike a balance between setting the direction for change and supporting local problem solving efforts. The goal of policy should be to enable reform, not to regulate and prescribe it.
4. *Create feedback mechanisms.* Policy makers need to hear from schools and school districts, especially those in high poverty and minority areas. Avoid sole reliance on official representatives of interest groups.
5. *Make mid course corrections.* Revisit policies as new information comes to light, and make adjustments when they are needed.
6. *Focus attention on education as a public good.* Emphasize the critical role of public schools in our democracy. Shift the debate away from laying blame for school shortcomings and toward a shared effort to make things better. This change will help keep small but vocal interest groups, individual horror stories, and political infighting from derailing the goals of reform.

Perhaps the key to science/math curriculum content change is the linkage to broader political priorities. The alleged link between improved math/science/technical education and future U.S. economic growth has fueled much of the political interest and increased finance to date.

ACKNOWLEDGMENTS

This paper is a revised version of a paper prepared for the Wisconsin Center for Education Research and the National Science Foundation for publication as an NISE Research Monograph by the National Institute for Science Education, University of Wisconsin-Madison.

REFERENCES

American Association for the Advancement of Science. (1990). *Project 2061: Science for all Americans.* Washington, DC: Author.

Archbald, D. A., & Newman, F. (1988). *Beyond standardized testing: Assessing authentic academic achievement in secondary school.* Reston, VA: National Association of Secondary School Principals.

Arocha, Z. (1993). The religious right's march into public school governance. *The School Administrator, 50* (9), 8-15.

Ball, D. L. (1992). *Implementing the NCTM standards: Hopes and hurdles.* (Issue Paper 92-2). East Lansing, MI: Michigan State University, National Center for Research on Teacher Learning.

Ballen, J. (1995, June). *Agendas, structure, and strategies of conservative Christian organizations.* Unpublished manuscript. Stanford University, School of Education.

Barth, P. (1993, October). The education holdup. *Education Week, 39*, 43.

Boyd, W. (1994). *Citizens for Excellence in Education: Mission statement and objectives of NACE/CEE.* University Park, PA: Pennsylvania State University, School of Education.

Brockett, D. (1993). Outcome based education faces strong opposition. *School Board News, 13* (10), 1-8.

Carlson, C. G. (1995). The metamorphosis of mathematics education. *Focus, 27.* Princeton, NJ: Education Testing Service.

Collins, A. (1995). National science education standards in the U.S.: A process and product. *Studies in Education Science, 26*, 7-37.

Cohen, M. (1983). Instructional management and social conditions in effective schools. In A. Odden & L. D. Webb (Eds.), *School finance and school improvement linkage for the 1980s.* Cambridge, MA: Ballinger.

Cuban, L. (1984). School reform by remote control: SB813 in California. *Phi Delta Kappan, 66*, 213-215.

Dow, P. B. (1991). *Schoolhouse politics: Lessons from the Sputnik Era.* Cambridge, MA: Harvard University Press.

Elmore, R. F. (1990). *Restructuring schools: The next generation of educational reform.* San Francisco: Jossey-Bass.

Elmore, R. F. (1993). *The development and implementation of large-scale curriculum reforms.* Background paper prepared for the American Association for the Advancement of Science, available from AAAS.

Elmore, R. F., & Fuhrman, S. (1994). *The governance of the curriculum.* Alexandria, VA: Association for Supervision and Curriculum Development.

Frederickson, N. (1984). The real test bias: Influences of testing on teaching and learning. *American Psychologist, 39*, 193-202.

Fuhrman, S. H. (1993). Politics of coherence. In S. H. Fuhrman (Ed.), *Designing coherent policy: Improving the system* (pp. 1-34). San Francisco: Jossey-Bass.

Grove v. Mead School District, 753 F.2d 1528 (9th Cir., 1985), *cert. denied*, 474 U.S. 826 (1985).

Hazelwood School District v. Kuhlmeier, 795 F.2d 1368 (8th Cir. 1986), rev'd on other grounds, 484 U.S. 260 (1988).

Kirst, M. W. (1994, August). The politics of nationalizing curricular content. *American Journal of Education, 102*, 383-393.

Kirst, M. W., Anhalt, B., & Marine, R. (1997, Winter). Politics of science education standards. *The Elementary School Journal, 96.*

Kirst, M. W., & Kelley, C. (1993, April). *Positive impacts of reform efforts in the 1980s: A Nation at Risk ten years later.* Paper prepared for a symposium. Berkeley, CA: University of California.

Kirst, M. W., & Massell, D. (1994, February). Determining national content standards. *Education and Urban Society, 26*, 107-117.

Koretz, D. M., Madaus, G., Haertel, E., & Beaton, A. (1992). *National educational standards and testing: A response to the recommendations of the Na tional Council on Educational Standards and Testing.* Santa Monica, CA: RAND.

Lindblom, C., & Braybrooke, D. (1963). *A strategy of decision.* New York: Free Press.

Marsh, D. D., & Odden, A. R. (1991). Implementation of the California mathematics and science curriculum frameworks. In A. R. Odden (Ed.), *Education policy implementation* (pp. 219-240). Albany: State University of New York Press.

Marshall, C., Mitchell, D., & Wirt, F. (1989). *Culture and educational policy in the American states.* New York: Falmer.

Marzano, R. J. (1993-1994). When two world views collide. *Educational Leadership, 15* (4), 18-19.

Massell, D. (1994a). Setting standards in mathematics and social studies. *Education and Urban Society, 26*, 118-140.

Massell, D. (1994b). Three challenges for national content standards. *Education and Urban Society, 26*, 185-195.

Massell, D., & Fuhrman, S. (1993, November). *Ten years of education reform: Overview with four case studies.* New Brunswick, NJ: Consortium for Policy Research in Education.

Massell, D., & Goertz, M. (1993, October). *2061 POLICY Blueprint.* Unpublished manuscript prepared for the American Association for the Advancement of Science, Washington, DC.

Massell, D., & Kirst, M. (Eds.) (1994). Setting national content standards [Special issue]. *Education and Urban Society, 26* (2), 1-2.

McCarthy, M. (1993). Challenges to the public school curriculum: New targets and strategies. *Phi Delta Kappan, 75*, 55-60.

McLaughlin, M. W. (1991). The RAND change agent study: Ten years later. In A. R. Odden (Ed.), *Education policy implementation* (pp. 143-155). Albany: State University of New York Press.

Mitchell, D. (1989). *Culture and educational policy.* Philadelphia: Falmer.

Mozert v. Hawkins County Public Schools. 827 F.2d 1058 (6th Cir. 1987), cert. denied, 484 U.S. 1066 (1988).

Myers, M. (1994). Problems and issues facing the national standards project in English. *Education & Urban Society, 26*, 141-157.

National Commission on Excellence in Education. (1983). *A nation at risk: The imperative for educational reform.* Washington, DC: U.S. Government Printing Office.

National Council of Teachers of Mathematics (1989). *Curriculum and evaluation standards for school mathematics.* Reston, VA: Author.

National Council on Educational Standards and Testing (1992). *Raising standards for American education.* Washington, DC: Government Printing Office.

O'Day, J. A., & Smith, M. S. (1993). Systemic reform and educational opportunity. In S. Fuhrman (Ed.), *Designing coherent education policy: Improving the system* (pp. 250-312). San Francisco: Jossey-Bass.

Olson, L. (1993, December). Who's afraid of OBE? *Education Week, 12*, pp. 25-27.

Page, L. (1995). *Making the grade: What Goals 2000 means to our schools.* Colorado Springs, CO: Focus on the Family.

People for the American Way. (1994). *Attacks on the freedom to learn [1993-1994] report.* Washington, DC: PAW.

Pence, P. & Petrosky, A. (1992, April). *Defining performance standards and developing an assessment for accomplished English language arts teaching of young adolescents.* Paper presented at the annual meeting of the National Council on Measurement in Education, San Francisco.

Purkey, S. C. & Smith, M. (1983). School reform: The district policy implications of the effective schools literature. *Elementary School Journal, 85* (4), 358-389.

Ravitch, D. (1995). *National standards in American education.* Washington, DC: Brookings Institute.

Smith, M., & O'Day, J. (1991). Systemic school reform. In S. Fuhrman & B. Malen (Eds.), *The politics of curriculum and testing* (pp. 223-267). New York: Taylor & Francis.

Smith v. School Commissioners of Mobile County, 862, F. 2d 1517 (11th Cir. 1989).

Sykes, G. & Plastrik, P. (1992). *Standard-setting as educational reform.* A paper prepared for the National Council on Accreditation of Teacher Education, Washington, DC.

Tyack, D., & Cuban, L. (1996). *Tinkering toward utopia.* Cambridge, MA: Hanan.

Tyson-Bernstein, H. (1988). The academy's contribution to the impoverishment of American textbooks. *Phi Delta Kappan, 70*, 193-198.

Virgil v. School Board of Columbia County, Florida, 862 F.2d 684 (11th Cir. 1989).

Whitty, G. (1984). The privatization of education. *Educational Leadership, 41* (7), 51-54.

Wirt, F., & Kirst, M. (1975). *Political and social foundations of education.* Berkeley, CA: McCutchan.

Wirt, F., & Kirst, M. (1992). *Schools in Conflict.* Berkeley, CA: McCutchan.

Yee, G. & Kirst, M. (1994). Lessons from the new science curriculum of the 1950's and 1960s. *Education and Urban Society, 26* (2), 158-171.

Zahorchak, G., & Boyd, W. (1994). *The politics of outcome based education in Pennsylvania.* Paper prepared for Annual Conference of the University Council for Educational Administrtation in Philadelphia, October.

Zykan v. Warsaw Community School Corporation, 631 F.2d 1300 (7th Cir. 1980).

A CLASSROOM PERSPECTIVE ON THE POLITICS OF CURRICULUM STANDARDS

Arthur E. Lehr

In their paper in this book, Kirst and Bird present an extensive overview of how policy makers outside of school classrooms deal with the politics of devising curriculum standards, but they say little about what happens when policy makers' decisions reach those classrooms. Because the classroom is the level at which curriculum standards are most important, a classroom perspective seems appropriate. In their discussion of sources that influence the determination of what is worthy to be included in the curriculum, Kirst and Bird refer to appeals to tradition, to science, to community, and to individual judgment, but they include no appeal to the setting in which curriculum standards apply. Perhaps an appeal to the classroom situation should be included to add the consideration of what is possible to the mosaic of sources and influences concerned with what is desirable.

Considering classroom situations in the process of devising curriculum standards calls for some thought to be given to the role teachers presently play in determining what is worthy of presentation to students. Despite the

Advances in Educational Administration, Volume 5, pages 133-144.

ISBN: 0-7623-0081-7

hierarchical structure of school systems, it is likely that teachers have the last word on what is worth teaching. Teachers have the opportunity to set standards in their classrooms every day based on what they find can be achieved, and these standards may or may not relate to those set by national, state, or district policy makers and officials. Given their present, nearly non-existent level of participation in formulating curriculum standards outside of the classroom, teachers may tend to regard standards received from outside sources more as annoying encumbrances than as inspiring guides to classroom practice.

LINKING CURRICULUM STANDARDS AND CLASSROOMS

Although Kirst and Bird suggest that the key to curriculum change is linkage to broader political priorities, a more effective alternative might be to refocus that change on the classroom by relocating the classroom teacher as a participant in determining curriculum standards and by establishing a direct linkage between curriculum expectations and classroom situations. Elmore (1996) affirms the need for a direct connection between the classroom setting and standardized achievement expectations. He proposes a "reciprocal accountability for capacity and performance" (p. 26) in which every increased performance demand carries with it "an equal reciprocal obligation on the part of the party making that demand to provide the capacity to meet it" (p. 27). The sharing of responsibility for meeting standards, however, may not fit well with the political priorities of many policy makers. It is one thing to establish optimal student performance expectations on the basis of public demand, but quite another to relate such expectations to actual student performance possibilities in the classroom.

Under Elmore's proposal, every expectation for student academic performance to meet curriculum standards would be accompanied by an obligation on the part of policy makers who set the standards to supply whatever resources and support are needed to make possible the achievement of those standards. This would require some radical changes. Elmore points out that policy makers would have to be prepared to "create the capacity to meet the demands they lay on teachers and principals" (p. 27). The credibility of policy makers would depend on their ability to find resources to implement solutions to the problems they identify. The distance between policy makers and teachers would necessarily lessen. And, possibilities for actual student achievement in real classroom situations would become a primary consideration in setting curriculum standards.

FORMULATING STANDARDS OUTSIDE OF THE CLASSROOM

Kirst and Bird disapprove of current procedures for making decisions about curriculum standards, seeing them as attempts to satisfy major interest groups through vague language and to avoid conflict by making only minimal changes. They are most critical of decision makers who use what Lindblom and Braybrooke (1963) call disjointed incrementalism, a strategy that simplifies complex problems and produces only marginal changes in policy. That criticism is well deserved in the view of many teachers who find that curriculum policy decisions usually are not made in or near classrooms and seldom consider the circumstances of students in those classrooms.

Support for the marginal incrementalism that Kirst and Bird criticize may be based more on easing the burden of maintaining the bureaucratic system of education than on addressing the needs of the classroom. The problem with incrementalism in decision making is that it does more to perpetuate the status quo than to bring about improvement (Peters, 1996), and the problem with making decisions to improve curriculum standards outside of the school setting is that they have too little to do with improving instruction. Maintenance of the status quo, according to Maeroff (1993), serves the purposes of school bureaucracy and is applauded by some administrators because it simplifies their jobs. That, however, does not address the concerns of instruction.

When schools function effectively, instruction happens in the classroom. Meeting curriculum standards, however, is a concern more often linked with the outcome, than with the input of classroom instruction. It is important to consider the results of instruction, but it is also important to consider that the quality of classroom performance may depend more on what the teacher and the students are able to bring to the experience than what they are able to carry away. Although policy makers generally assume that teachers are key actors in applying curriculum standards to instruction, they do not seem to grasp that teachers must also be active participants in the process of developing curriculum standards.

Elmore, Peterson and McCarthey (1996) describe the struggle of a fourth-grade teacher to meet a National Curriculum Standard set by the National Council of Teachers of Mathematics requiring that K-4 children understand place-value concepts. The standard was adopted by the school district as a measure of the outcome of elementary teachers' work with students. The district, however, gave teachers neither the time nor the opportunity to develop an understanding of the place-value concepts set forth in the standard. Although the teacher in the Elmore, Peterson, and McCarthey study attempted to meet the standard through a new activity-based approach, she later discovered that she had not presented the underlying logic of place value that the standard required (pp. 106-107). In this instance, the standard was installed as a measure of the outcome of classroom instruction, without becoming a functional part of classroom teachers' instructional repertoire. Teachers were bypassed in the process of establishing the standard, and instruction suffered. This case may be typical of what happens when teachers are treated as non-participants in determining curriculum standards.

Teachers may be faulted for failure to apply a standard properly, but given a situation in which they have no role in creating the standard and little or no contact with its makers, blame may conceivably be shared when instructional efforts go awry. As Elmore, Peterson, and McCarthey point out:

> The issue of how to teach place value with understanding continues to be problematic for both local communities of mathematics teachers and the national community of mathematics-education researchers. The two communities continue to grapple with this issue, but independently, and the problems of each community remain relatively invisible to the other (p. 108).

So long as those who set curriculum standards work separately from those who apply the standards, confusion and misdirection are more than likely to occur. A school system may rightly hold classroom teachers accountable for meeting standards, but may also need to consider what resources and opportunities must be made available to those teachers to fulfill their obligations.

Curriculum standards set uniform expectations for learning achievement, and in so doing, suggest an assumption of uniformity in teaching practice as well as student ability. That sort of consistency may exist only in the minds of policy makers and school officials who see good schooling as a matter of efficient management, routine supervision, and an appropriate division of labor. If school decision makers and officials adopt such a bureaucratic attitude, they may regard curriculum standards as no more than quality control measures necessary to ensure a satisfactory product. This view may further assume that teaching can be facilely defined in terms of what teachers are supposed to do and can be subsequently monitored by relatively simple evaluative routines. Within so narrow a perspective, school policy makers and officials may be content to fault teachers when standards are not met and may not feel obliged to inquire about what should be added to the classroom to help make teachers more productive.

Darling-Hammond (1990) presents substantial criticism of decisions concerning instruction that are formulated elsewhere and then handed down to classroom teachers. She explains that "effective teaching is not routine, students are not passive, and questions of practice are not simple, predictable, or standardized" (p. 30). Facilities and resources found in schools and classrooms across the country are so extraordinarily diverse as to suggest that standardized curriculum expectations are impractical, if not impossible to achieve. Decision makers external to the school setting—state legislatures, boards of education, administrative officers, and national subject matter associations—tend, however, to partition notions of what ought to be from what actually is. Curriculum standards that play well in local, state, and national political arenas are not necessarily feasible in real school classrooms.

CLASSROOM REALITIES

Kirst and Bird note that those who decide what to teach are torn between the needs of society and the needs of the individual; however, another tear in the fabric of decision making may exist between the needs of the classroom and the needs of the school organization. Because the classroom situation exists within the school situation, needs identified by the school organization may intrude upon implementation of curriculum standards. One such need may result from the inadequacy of school facilities. For example, science curriculum standards may call for a high level of student proficiency in laboratory procedures. To the science teacher with thirty-four students in a classroom equipped with twenty functional lab stations, those standards take on an ethereal quality. An overcrowded and underfunded school may "need" to assign thirty-four students to each science class, but doing so makes student achievement of standards extremely difficult, if not impossible.

Another example of the intrusion of organizational needs may result from district policies on promotion and retention. A high school English department, for example, may fully intend to meet writing curriculum standards, but may find that more than half of the students in composition courses are two or more grade levels behind in their writing skills. A school system that fears to damage student self-esteem, to offend parents, or to appear ineffectual may "need" to pass students from elementary or middle school to high school regardless of their academic progress, but doing so places nearly insurmountable obstacles in the path of teachers attempting to meet student achievement standards.

It is not difficult to imagine who is held accountable in the public domain when standards are not met in situations like those described above. "If the children haven't learned, the teacher hasn't taught" is still a mantra among the least knowledgeable and most political of school critics. Demands for accountability and high standards in education may be sound political moves, but they are not necessarily consonant with classroom realities and, therefore, may not be well received by students and teachers.

A consideration often overlooked in the push for accountability and high standards is the motivation of students. The emphasis on meeting high standards of academic performance may suggest to students that they are valued solely on the basis of their academic prowess. Noddings (1992) suggests that, "children are too often valued only for their academic achievement. They become resources" (p. 13). Teenagers, in particular, may not be impressed by the demands and decrees of educational and political leaders who stridently call for quality education and high expectations for all. Many teenagers today see school as a kind of "cattle ramp" where they are part of a herd, pushed and prodded along without any recognition or concern for their interests or goals. They tend to conclude that the primary focus of adult interest in them is discovering how well they measure up to academic standards mysteriously set by people whom they have never met.

Noddings (1992) explores the reaction of many public school students to the contemporary hue and cry for improving education by raising standards, noting that they may be particularly sensitive to the frequent criticism of adults who find them wanting:

> Lots of young people see through today's educational slogans. We preach constantly that "all children can learn"; we even suggest strongly that they all can learn anything the school has to offer if they are taught well and they try. If they don't try, they are made to feel like traitors, even though they might work very hard at tasks over which they have some control and choice. Thus, despite our determined optimism and insistent everyone-can-do-it, students complain, "They don't care!" They suspect that we want their success for our own purposes, to advance our own records, and too often they are right (p. 13).

The persistent demands for student achievement of high performance standards by the school system may, indeed, produce an effect opposite from that intended, especially among students whose prior experiences have been neither rewarding nor successful. Constant attention to the measure of their achievement may contribute to students' sense of powerlessness, and the threat of imminent failure may lead them to resent and reject the intrusion of the school into their lives.

Teachers have also felt some resentment of the intrusion of the school organization—its demands and mandates—into classroom instruction, and their response has varied from quiescent acceptance to active resistance. Teachers have always found ways to deal with obstacles placed in their paths by the school organization. In school settings where professional development is a low priority and mostly confined to an occasional presentation by an inspirational speaker or visiting expert, teachers find time before or after school, on weekends, and during the summer to consult with their peers, take college courses, or inform themselves through reading and research. In schools where facilities, classroom equipment, and instructional materials are inadequate, teachers write grant proposals, sponsor fundraisers, or spend their own money to fill the gap. In communities where school policies result more from political than educational interests, teachers organize and become active in local politics to advance the needs agenda of their classrooms. These ad hoc efforts by teachers are, of course, band-aids for the underlying problems that beset school classrooms. Teachers spend a great deal of time treating symptoms because they are permitted little or no role in working out cures for classroom ills. Teachers must have a far greater role in planning and preparing the objectives, the content, and the means of instruction if they are to respond successfully to the harsh and frequent criticisms leveled at today's schools.

OUTSIDE POLICIES—INSIDE SKEPTICISMS

Kirst and Bird assert that federal and state policies tend to support use of low-level curricula geared to minimum competency levels (as defined by standardized tests), rather than to support development of high-level skills, and that teachers tend to take the easy way out by preparing students for low-level test expectations. These criticisms may be accurate, but there is another perspective to consider. Teaching and learning languish when local school boards and administrators lower standards to hold down student failure rates so that their schools do not appear to waste resources. As Shedd and Bacharach (1991) point out, "school efficiency tends to be defined in terms of providing a fixed and minimum level of service at the lowest possible costs, and setting minimal 'product specifications' is the easiest way to guarantee that all or most teachers and students can meet them" (p. 53). Sometimes more is done to maintain the face of public education than to support its classroom practices.

In examining the results of top-down policy making and its effects on schools, Kirst and Bird explain:

> Educators have tended to be highly skeptical about policymakers' ability to develop ambitious and challenging standards for student achievement that can actually be carried out in the classroom. Their skepticism is based on the fact that it is easiest for politicians to set standards at levels that the current capacity can readily achieve (pp. 107-132).

Kirst and Bird are presumably accurate in observing that educators have doubted policy makers' ability to develop meaningful achievement standards and subsequently have rejected top-down reform, preferring school-based efforts founded on local need.

Skepticism about top-down policies, not unusual among classroom teachers, may be an adverse reaction to policy makers' traditional assumption that because schools are government agencies, education policies are properly made only at the top of the system. In situations where policy makers simply announce curriculum standards without advance preparation or support, teachers may become suspicious and even hostile, not only toward the source of the imposed performance expectations, but also toward the standards themselves. Elmore (1996) asserts that teachers are suspicious of top-down policy systems, "not because they don't believe that they are ultimately accountable for what students learn, but because they believe that most administrators and policymakers don't have any idea of the practical demands they are imposing on teachers and students through their actions" (p. 27). When teachers have no intellectual input or investment in curriculum standards, they are generally not inclined to treat them kindly and may react negatively.

Policy makers, especially those outside of the schools, sometimes express surprise and dismay that their decisions are not well received by teachers.

Surprise and consternation are understandable in policy makers who are so disconnected from the school setting as to assume that the classroom situation will simply accommodate itself to meet their expectations, whatever they may be. Because they are disconnected from classroom realities, policy makers may, in fact, harbor some ingenuous and pessimistic notions about how schools and teachers work. Darling-Hammond (1990) argues that decision makers who advocate and conduct top-down policy procedures assume that whatever needs to be accomplished in education can be accomplished through "more precise regulation of educational or management processes" (p. 27). O'Neil (1991) suggests that policy makers who employ top-down procedures assume that teachers must be directed and supervised in their work because they are "basically lazy and slothful" (p. 3). Teachers are usually offended by such unreasonable assumptions, although they understand them to be common among policy makers who are out of touch with the everyday world of the public schools.

Teachers also may not respond well to top-down policies because they take note of the considerable distance invariably maintained between school policy makers and school classrooms. Elmore (1996) offers the example of a primary grade teacher who summed up the problem of carrying out the decisions of policy makers who demand teacher accountability but have little or no classroom contact. As she put it:

> Just once, I'd like the guys in the suits to come into my classroom and get down on their hands and knees with me and see what it's like to actually teach this stuff. Then, I think we could have an intelligent conversation about accountability (p. 27).

If Rodney Dangerfield were a teacher, he might simply say that teachers' react as they do because they "don't get no respect."

BRIDGING THE GAP BETWEEN CLASSROOMS AND POLICY MAKERS

Because past school reform efforts have not been successful, policy makers today, Kirst and Bird observe, are well aware of problems that follow if plans for change are not widely shared at the local level and that, subsequently, many standards projects now attempt to include professional educators, community members, and others with an interest in curriculum standards. Kirst and Bird further note that efforts to reduce tensions and to bridge the gap between policy makers and classroom teachers presently include attempts to give teachers and local school districts some discretion in how to go about implementing standards. Standards, as Kirst and Bird describe them, serve as "a sort of flag that reflects the valued goals around which educators can rally and decide for themselves how the objectives of the standards will be accomplished in their schools and classrooms." Kirst and Bird also warn, however, that standards may become so flexible that they no longer provide viable goals.

Allowing teachers discretion in the application of standards may indeed be a good and necessary idea, but a more important pliancy for policy makers to consider may be flexibility in allocating resources to meet student needs and in supporting teaching methods, materials, and processes, as these are areas in which policy makers tend to be least flexible. The standardization of input processes—budgets, personnel assignments, and instructional materials—is, in Darling-Hammond's (1990) view, a symptom of counterproductivity in the operation of school systems. Uniform inputs into a system in which no two people, places, or needs are alike may be less than rational. If that is so, then how rational is the expectation of uniform outcomes? Setting standards and being flexible by avoiding teaching practice prescriptions to accomplish those standards may help relieve tensions between policy makers and teachers, but setting standards and being inflexible in providing what is needed to accomplish them can only generate more tensions.

Kirst and Bird cite dilemmas and trade-offs faced by policy makers and educators in setting standards, including full participation of all interest groups versus process efficiency, broad consensus versus specific progress, subject area versus interdisciplinary content, insufficient instructional guidance versus prescriptive pedagogy, and subject matter specialists versus practitioners. Another dilemma may be found in expectations set by standards versus classroom possibilities limited by local situations and resources. If teachers apply uniform curriculum standards of achievement without considering the diversity of individual students and the complex situational problems of the classroom, they may be sanctioned for producing a high failure rate that brings their competence into question. If they adapt standards to their classroom situations and the individual needs of their students, they may be sanctioned for failure to enforce uniform achievement standards. This dilemma is the inevitable result of limiting or excluding the meaningful participation of the majority of teachers in setting student achievement standards.

Using categories adapted from Marshall, Mitchell, and Wirt (1989), Kirst and Bird chart the participants who are influential in determining state and local policies from the insiders to the sometimes players, but do not include classroom teachers anywhere in their rankings. Guidelines recommended by Kirst and Bird for pursuing the development of standards urge professional development to "help redefine the job of teaching," call for balance between top-down and bottom-up strategies to enable reform, advocate feedback mechanisms to allow policy makers to "hear from schools and school districts," and support mid-course corrections to "make adjustments when they are needed." These measures appear mainly to support conformity and compliance with policy requirements from those who implement them.

Under the guidelines offered by Kirst and Bird, classroom teachers are not even on the periphery of participation in formulating standards. Kirst and Bird do acknowledge that teachers are involved to the extent that they need help in changing their practice to meet new expectations, that higher level change policies support their efforts to change, and that they respond to policy makers' need to hear how things are going. In terms of influencing curriculum standards development, however, classroom teachers appear to be out of the loop.

It may be argued that teachers have influence through their representatives in state or national subject area organizations. This might be a valid assumption if the majority of classroom teachers were active participants in those organizations. Other than occasional trips to regional, state, or national conferences supported by very limited local funding and periodical publications from subject area groups, however, most teachers have little or no contact with such organizations. They typically have even less contact with the other players Kirst and Bird identify as influential in standards development, being far more likely to receive from them announcements of decisions than inquiries for information or advice.

Kirst and Bird also chart the key players for standards setting at the district level, including school boards, teachers unions, administrators, curriculum supervisors, parent groups, and department chairs. Other than union activists and the handful of teachers who serve as curriculum supervisors or chair departments, classroom teachers are more likely to be spectators than participants in local decision making. School boards typically demand and receive reports on teaching activities, but usually filter them through the district hierarchy. Dialogs may occur between school boards, administrators, and union representatives, but they tend to be formally structured by contractual agreements limiting them to concerns involving pay and working conditions. Parent groups may have strong ties with teachers, but their participation in decision making, like that of most teachers, tends to be advisory at best.

Early in their paper, Kirst and Bird note that "those matters that deserve first priority in the curriculum are to be determined by the community, either directly, via its representatives, or by studies of the community." Are teachers a viable part of the local community that determines curriculum priorities? Other than the voice of their union representatives and that of an occasional, outspoken individual, it appears that teachers typically live on the fringe of community participation in school decision making and, therefore, may seldom be heard. This barely marginal level of participation may be related to the isolation that many teachers, especially those at the secondary level, experience in the regimen of their daily work.

It may be that the professional isolation of many teachers in the traditional structure of schools has led to or at least contributed to their acceptance of a somewhat less than marginal role in determining curriculum standards at the national, state, or local levels. Teachers who are segregated into classrooms and subject area departments and who have only minimal contact with their peers during the working day have little opportunity to see themselves as school constituents. Attempts at reaching or even exploring schoolwide consensus on what should be taught are rare in most schools that are typically driven by an inflexible time schedule that disallows collaboration as a regular feature of the work environment.

When teachers become separate, unconnected entities, according to Inger (1993), they may begin to operate as individuals who feel that what the school does or what other teachers do is none of their business. Johnson (1990) notes that because teachers spend most of their working day with students and have so little contact with their peers, they can become quite detached from the school program and set about "devising curricula on their own, ignoring the plans and practices of their counterparts in other classrooms" (p. 148), and contentedly working out their own interpretations of what and how to teach. Akin (1990) calls this "teacher self-actualization" (p. 6), in which teachers devise lesson plans and select and prepare instructional materials without knowing how or if their efforts complement or conflict with the rest of the school program. The isolation of teachers within the work environment may have the net effect of producing a school curriculum that has a plethora of aims and functions, but no central focus and no unified vision of educational purpose. The imposition of curriculum standards is likely to have little effect in a setting so fragmented and among teachers so sequestered from a central role in determining what should be taught.

CONCLUSION

Teachers must have a central role in determining curriculum standards if those standards are to have any real meaning in the classroom. Policy makers must recognize that their responsibilities reach far beyond merely setting expectations and demanding accountability. Reciprocity between policy makers who determine curriculum standards and classrooms where those standards are to be applied is essential to solving the problems of school settings and achieving the goals of public education.

If policy makers accept responsibility for the impact of their decisions on schools and teachers and students, the classroom situation may then become the starting point for setting curriculum standards and classroom teachers and students may finally have a place in the inner circle of influential participants in the process. At present, however, this ideal of meaningful participation may be far beyond what most policy makers will ideologically embrace. The "quick fix" of unilaterally raising standards and imposing new accountability systems to deal with perceived shortcomings in education still holds the greatest appeal for legislators and other policy makers who have no immediate ties to the reality of the public school.

REFERENCES

Akin, J. (1990). *When the restructuring bandwagon reaches high school, will site culture bar the door?* ERIC Document Reproduction Service No. ED 352 728.

Darling-Hammond, L. (1990). Teacher professionalism: Why and how? In A. Lieberman (Ed.), *Schools as collaborative cultures: Creating the future now* (pp. 25-51). New York: Falmer Press.

Elmore, R. F. (1996). *Accountability in local school districts: Learning to do the right things.* Paper prepared for the First Edwin O'Leary Symposium on Financial Management at the University of Illinois at Urbana-Champaign, October 16-17.

Elmore, R. F., Peterson, P. L., & McCarthey, S. J. (1996). *Restructuring in the classroom: Teaching, learning and school organization.* San Francisco: Jossey-Bass.

Inger, M. (1993). Teacher collaboration in urban secondary schools. *ERIC/CUE Digest,93.* New York: ERIC Clearinghouse on Urban Education.

Johnson, S. M. (1990). *Teachers at work: Achieving success in our schools.* New York: Basic Books.

Lindblom, C., & Braybrooke, D. (1963). *A strategy of decision.* New York: Free Press.

Maeroff, G. I. (1993). *Team building for school change: Equipping teachers for new roles.* New York: Teachers College Press.

Marshall, C., Mitchell, D., & Wirt, F. (1989). *Culture and educational policy in the American states.* New York: Falmer.

Noddings, N. (1992). *The challenge to care in schools: An alternative approach to education.* New York: Teachers College Press, Columbia University.

O'Neil, J. (1991, December). National push for school reform, accountability to influence education's future agenda. *ASCD Update.* Alexandria, VA: Association for Supervision and Curriculum Development.

Peters, P. B. (1996). *American public policy: Promise and performance* (4th ed.). Chatham, NJ: Chatham House.

Shedd, J. B., & Bacharach, S. B. (1991). *Tangled hierarchies: Teachers as professionals and the management of schools.* San Francisco: Jossey-Bass.

ASSESSMENT LINKAGES WITH INSTRUCTION AND PROFESSIONAL STAFF DEVELOPMENT

Delwyn L. Harnisch

Reform of elementary and secondary education has been a hallmark of U.S. education policy of the last fifteen years. The reform efforts were sparked by a number of reports published in the early 1980s decrying the quality of our nation's schools. *A Nation at Risk* by the National Commission on Excellence in Education (1983) was a prominent force during this time frame. Other groups specifically documented inadequate student achievement and preparation in mathematics and sciences (National Academy of Science, 1982; National Science Board, 1983).

In their paper in this book, Kirst and Bird present an overview of the major national reforms in the 1980s and direct our attention to how these reforms are motivated by political and social challenges and aimed at all students. The titles of these major efforts demonstrate that the goal for the 1990s is science and mathematics for all. Two examples are the American Association for the Advancement of Science's (AAAS) *Science for All Americans* produced by Project 2061 (AAAS, 1989; AAAS, 1995) and the Mathematical Sciences

Advances in Educational Administration, Volume 5, pages 145-164.

ISBN: 0-7623-0081-7

Education Board's (1989) *Everybody Counts*, associated with the mathematics standards developed by the National Council of Teacher of Mathematics (NCTM). Kirst and Bird describe a standards based reform that focuses on curriculum standards with a strong motivation to examine the assessment of student achievement. This paper examines the standards and a rethinking of assessment for mathematics in this reform context.

STANDARDS

The general perception has been that U.S. students perform poorly in mathematics compared to their counterparts in other countries. Standards were developed to describe what students should know and be able to do in a given curriculum field. Standards are intended to present a vision of curriculum and teaching in a given field and to provide criteria for state and local school personnel so they can decide on the specifics of what should be taught, how it should be taught, and how student learning should be assessed. What is described in these reforms are nationally developed standards, but not national standardization.

The use of the term "standards" is a powerful concept because it encompasses several meanings. NCTM has used the term as a rallying phrase for professionals in the field. NCTM defines standards as serving three purposes: (a) a minimal criteria for quality (intended to check on curricular and textbook claims), (b) an expression of expectations of goals, and (c) a means for leading a group toward new goals (NCTM, 1989). In short, NCTM's standards publications represent statements about education in mathematics that exemplify thinking about best practice.

For the policy makers, the meaning of standards is often linked with accountability. In this context, the standards consist of a set of statements against which objects and activities are measured. The primary interest in standards is related to making schools, teachers, and students perform at certain level. An example of this use can be seen in the Illinois Goals Assessment Program (IGAP) where performance patterns are described in terms of the percentage of students meeting the standards of performance for various learning outcomes. Discussions of different meanings for standards are running rampant in school based settings and are likely to lead to controversy when the different meanings begin to clash.

Kirst and Bird describe the influence of the standards setting movement in the development of AAAS's Project 2061. This project, according to Kirst and Bird, represents an integration of both top down and bottom up reform strategies working together to change the educational system. Although the desirability of national standards was seldom addressed, the project was forced to respond to discussions surrounding the evolution of the national science standards developed by the National Research Council (NRC) (1996).

In short, the use of the term standard refers to the educational system on two dimensions: (a) how students can be brought, or can bring themselves, to meet the standard; and (b) an assessment of whether the standards have been met. What is common between these two points is an assessment of student achievement that is matched with the standards. Policy makers and reform sponsors often prefer assessments that are designed and controlled by sources external to the classroom over assessments by the classroom teacher. In the United States we have a long history of using objective tests for student placement at both lower and higher levels of education in elementary and secondary schools. During the past twenty-five years the amount of student testing has greatly increased because federally funded compensatory education programs have often required standardized testing to establish student eligibility.

In the 1960s, emphasis was given to developing curriculum materials. From their experience at that time, NCTM leaders involved with the development of curriculum standards recalled that just changing the curriculum was not sufficient to improve mathematics education. Given this background and the shortage of qualified mathematics teachers in the early 1980s, NCTM chose to focus its efforts on developing both curriculum standards and teaching standards (McLeod, Stake, Shappelle, Mellissionos & Gierl, 1996).

Acceptance of the standards based movement was not unanimous. Opposition to standards led reform at the national and state level has emerged (Ravitch, 1995). Some opponents expressed concerns with what they saw as an imposition of a particular set of attitudes and values; others discussed the unfairness of performance standards that did not include program and opportunity to learn standards for poor and minority students; and still others raised questions about the extent to which an assessment program could describe proficiency levels based on standards as measured through a national test or comparable state tests. Given these concerns, the long-range future of the standards movement in U.S. education is a matter for debate.

The NCTM standards documents (NCTM, 1989, 1991, 1995) and the Project 2061 Benchmarks (AAAS, 1993) served as primary sources for many of the implemented state curriculum guidelines. For example, thirty-two states have frameworks for mathematics and ten have combined science-mathematics frameworks (Blank & Pechman, 1995). The state guidelines vary greatly from state to state with a majority of the states implementing reforms in teaching and teacher preparation. Much legislation has been passed producing little effect. Policy makers, along with legislators, now understand that band-aid approaches to reform will not provide a cure for a system that was designed for America's past. What has been learned is that the most important school related factor in determining student achievement is the *expertise of the teacher*.

In his paper, Lehr provides a classroom perspective on the politics of curriculum standards and the importance of the role of the teacher in reform efforts. He points to the important role of teachers in educational reform and to the vital part they play in relating curriculum standards to models in action at the classroom level. I agree with Lehr's view on the importance of the classroom teacher in the current reform arena. They are critical agents in bridging the standards to the subject matter and to the classroom setting.

Lehr directs our attention to the need to involve teachers in the linkage of standards to the classroom context. He calls for the involvement of classroom teachers as partners in bringing about desired reform. Kirst and Bird describe how consensus on goals and support for standards depend on widespread dissemination, which is in turn assisted by enrolling many partners and collaborators in a joint effort.

WHAT KIND OF ASSESSMENT?

For the past several decades, the assessment of educational systems, schools, and teachers has evoked outcries from the general public and their legislative representatives. Assessment methods, based on standardized tests, have come under critical review in the educational community. Reform advocates have argued for alternative methods of educational assessment. A more balanced approach is needed to provide multiple indicators of student learning, including those supplied by standardized tests. Assessments have the capacity to describe the level of student proficiency and to show how well school systems are meeting the call for higher standards of excellence.

To measure student achievement with the best measures already available is likely to mean inclusion of a highly refined system of scholastic aptitudes. Stake (1995) notes that psychometricians have developed an excellent set of scholastic aptitudes that correlate highly with teacher grades and other measures of achievement, but technically do not measure achievement. Innovative programs, when viewed through summary aptitude measures, are likely to be judged as ineffective. Criterion based examinations are needed to be sensitive to the innovation. Program opponents, however, are likely to presume that the reformers' tests are biased toward their innovation. What measures of student achievement are used to describe a program's effectiveness? Changes in student achievement based on conventional examinations are often treated as the indicators of program quality. The standards based reform has focused on the reconceptualization of teaching and learning.

Assessment is a way of measuring what students know and of expressing what students should learn. For example, the role of mathematics in society has changed; therefore, mathematics education is changing, based on new national standards for curriculum and instruction. Assessment of mathematical understanding must also change to ensure consistency with the goals of education.

The Mathematical Sciences Education Board (1993) contends that three fundamental educational principles (content, learning, and equity) form the foundation of all assessment that supports effective education. In the content principle, assessment should reflect the mathematics that is most important for students to learn. With the learning principle, assessment should enhance mathematics learning and support good instructional practice. And, with the focus on the equity principle, we see how assessment should support every student's opportunity to learn important concepts in mathematics.

I believe that the principles of content, learning, and equity are important and contain a solid foundation to review the present reform efforts. In the United States to date, few assessments reflect any of these vital principles. To measure the progress made in our educational improvements and for educational reform to succeed, the principles of content, learning, and equity must be taken into account. These principles will be discussed within the framework for assessments matched with performances (the content principle), for assessments linked to instruction (the learning principle), and for assessment results based on the opportunity to learn (the equity principle).

In rethinking assessment in mathematics, test developers often acknowledge the importance of curricular and educational issues. Their concern, however, is usually about coverage, so they design tests by following check-off lists of mathematical topics (e.g., fractions, single-digit multiplication). This way of determining test content matched fairly well the old vision of mathematics instruction. In this view, students could look at little pieces of learning, add them up, and get the big picture of mathematics.

Today we recognize that students must learn to reason, create models, prove theorems, and argue points of view. Assessments must reflect this recognition by following closely the three principles of content, learning, and equity. Students cannot get at this kind of deep understanding and use of mathematics by examining little pieces of learning. Assessments that are appropriately rich in breadth and depth provide opportunities for students to demonstrate their deep mathematical understanding. Such a common vision must be shared between mathematics education and mathematics assessment.

The assessment of student achievement must change because today's students face a world that demands new knowledge and skills. In general, the students in the Twenty-first century need not only understand the basics, but also be able to think critically, to analyze situations, and to make inferences. The agents associated with change in the learning goals for students and schools include changes in the skills and knowledge needed for success, changes in understanding how students learn, and changes in the relationship between assessment and instruction. As a result of these changing conditions, we need to change our assessment strategies to match assessment design and content to the goals and purposes of assessment (Harnisch & Mabry, 1993; Harnisch, 1994a).

The pressures to change mathematics education reflect society's disappointment with the lack of interest and accomplishment of so many students in today's schools. In the background of public debate is the steady criticism that school mathematics is out of step with today's world and is neither well taught nor well learned. Concern is focused on the skills that are needed in an information age. Mathematics, for Twenty-first century use, must be made meaningful to students if it is to be learned, retained, and used. There is also a growing recognition that in this information age, all students should learn more and more useful mathematics.

Students are expected to have the ability to access, interpret, analyze, and use information for making decisions. In response to these changing competencies needed in the workplace, curriculum content standards and higher expectation levels for student skills and behavior have evolved. National and state level content standards have been developed in mathematics, science, geography, and history and have been synthesized in the Systematic Identification and Articulation of Content Standards and Benchmarks (Marzano & Kendall, 1995). At the center of many of these school/educational improvement efforts is the policy makers' hope that assessment changes will motivate teachers and schools to do things differently (Harnisch, 1994a; Linn, 1983, Madaus, 1985). In short, what is needed in assessment reform is a process whereby we set more appropriate targets for students, focus the professional development of teachers, encourage curriculum innovation, and improve instruction and instructional materials (Darling-Hammond & Wise, 1985).

ASSESSING STUDENT LEARNING THROUGH PERFORMANCE MEASUREMENTS

Recommendations from NCTM are very supportive of "performance testing." Performance testing often involves placing students in a problem situation where strategies and multiple sources of information are required, where contexts are somewhat realistic, and where students have the opportunity to work on solutions over a period of time. Performance assessment techniques include paper and pencil tasks (i.e., solve this story problem and explain how you solved it, or write a story about a typical day in Paul's life after you study the following graph that shows how he uses his time), naturally occurring events (i.e., observe a student's way of dealing with peer conflicts on the playground), longer term projects (i.e., collect and classify newspaper and magazine advertisements), portfolios (best work or growth and learning progress), demonstrations, experiments, oral presentations and dramatizations, and simulations and contrived situations.

Assessments should focus on tasks that elicit student performance on the meaning, process, and uses of mathematics. Mathematics considered important by the professional community must shape and define the content of assessment. What the mathematics teaching community must then do is choose appropriate tasks that emphasize connections within mathematics, embed mathematics in relevant real life problems, require students to communicate clearly their mathematical thinking, and facilitate the solving of challenging problems. Considerations of connections, communication, and challenging problems raise many complex issues that psychometricians are just beginning to research. As a result of this change, the issue of the nature of validity as a psychometric construct and what constitutes validation activities has been reopened for debate. (See *Educational Measurement: Issues and Practice*, Volume 16, Number 2, Summer 1997, for a special discussion on this issue.)

Interestingly, teachers today and in the past view the integrity of the classroom experience more with what students do in the classroom than with what they do on tests. Stake and Raizen (1997) note:

> Teachers have seldom expected external tests to capture what they were trying to do. Yet most teachers have come to accept standardized tests as important indicators of student progress and to see that preparation for those tests is one of their obligations. At the same time, innovative teachers have come to see performance tasks as an essential part of instruction, useful even if their inclusion in large-scale assessments and reporting of results remain problematic. Teachers and other educators committed to performance testing but also needing norm-referenced information will have to find ways to extend the testing period (p. 133).

A common belief held by educators and policy makers is that what is on the test is what gets taught and that the assessment format influences the instructional materials and format of instruction (O'Day & Smith, 1993). Assessment can be described as the guidance system of education, just as standards are the guidance system of reform. Assessments help teachers and parents to determine what students understand and what they need to learn. Assessments have the ability to share a picture of how well students are learning and how well school systems are meeting the call for higher education standards.

Assessments used today, such as standardized multiple choice tests, have reinforced the view that the mathematics curriculum should be constructed from lists of narrow, isolated skills that can be easily disassembled for appraisal. The assessments that are needed, however, are broader and more integrated with the new vision of school mathematics. Assessment should emphasize problem solving, critical thinking, and reasoning. In assessment, as in curriculum activities, students should build models that connect mathematics to complex, real world situations and regularly formulate problems on their own, not just solve those structured by others. Rather than forcing mathematics to fit assessment, assessment must be tailored to the mathematics that is important to learn.

Aligning Assessment With Instruction

Most educators and evaluation experts agree that there should be alignment between student assessment and instruction. An examination of a test should reveal the major learning targets that were covered by the instruction. The test results should identify the learning targets that are well understood and other targets that should have been taught. The traditional multiple choice and true-false assessments often test basic facts and skills in isolation without requiring students to apply what they have learned and can do in authentic types of problems. It has been noted that these traditional standardized tests do not match the newly written content standards (Wilson & Corbett, 1991). Shepherd (1991) argues that these tests often encourage instruction in less important skills and passive learning. What is needed are assessments that are based on current cognitive psychological models of learning, which contend that learners gain understanding when they construct their own cognitive maps of the interconnections among concepts and facts (Harnisch, 1995a).

Assessment results have the potential to be informative to students, teachers, parents, and policy makers. For example, results from assessments provide information that is used for assigning grades, evaluating a curriculum, and deciding whether to review multiplication facts. Classroom assessments provide information to teachers about critical aspects of students' performance and help teachers to adjust their instructional techniques accordingly (Harnisch & Switzer, 1991). Statewide tests (i.e., IGAP, 1995) provide information about the quality of instructional programs to parents, state and local education agencies, legislators, and policy makers.

Reformers often view assessment as much more than a report card. Some reformers have viewed assessment as the engine that propels reform by making education rather than the measurement the moving force in the design and development of new assessments. With new standards that are a visible target for all to see, it is felt that assessment can inform students, parents, and teachers about the real performance based meaning of the curriculum guidelines. Assessments provide not only measures of what students know, but also concrete illustrations of the important goals to which all teachers and students aspire.

Implications of the content principle also extend to the scoring and reporting of assessments. New assessments will require new kinds of scoring guides and ways of reporting student performance that more accurately reflect the richness and diversity of mathematical learning than do the typical single number scores of today (Harnisch, 1995). In short, accurate and useful reporting of assessment results that follow the content principle enables teachers, students, parents, and the public to understand why various assessment instruments are being used and how the results can be used as part of the school improvement process. To meet this goal, math educators must be involved in setting the goals for reporting, selecting strategies for reporting assessment results, and providing communication channels to students, parents, and the public that are appropriate for their level of understanding.

To be an effective part of the educational process, assessment should be seen as a linkage and an integral part of teaching and learning, rather than as the culmination of the process. Time spent on assessment will then contribute to the goal of improving the mathematics learning of all students. If assessment is going to support learning, then assessment tasks must provide genuine opportunities for all students to learn significant mathematics. Too often a sharp line is drawn between assessment and instruction. Teachers teach, then instruction stops and assessment occurs. In the past, for example, students' learning was often viewed as a passive process whereby students remembered what teachers told them to memorize. Consistent with this view, assessment has often been thought of as the end of learning. The student is assessed on material previously presented to determine if he or she remembers it. Earlier conceptions of the mathematics curriculum as a collection of fragmented knowledge led to assessment that reinforced the use of memorization as a principal learning strategy.

Today students make their own mathematics learning individually meaningful. Learning is a process of continually restructuring prior knowledge, not just adding to it. Good education provides opportunities for students to connect what is being learned with prior knowledge. Students know mathematics if they have developed the structures and meanings of its content for themselves.

If assessment is to support good instructional practice, then assessment and instruction must be better integrated than is commonly the case today. Assessment must enable students to construct new knowledge from what they know. The best way to provide opportunities for the construction of mathematical knowledge is through assessment tasks that resemble learning tasks in that they promote strategies such as analyzing data, drawing contrasts, and making connections. This can be done, for example, by basing assessment on a portfolio of work that the student has done as part of the regular instructional program, by integrating the use of scoring guides into instruction so that students will begin to internalize the standards against which the work will be evaluated, or by using two stage testing in which students have an extended opportunity to revise their initial responses to an assessment task.

Not only should all students learn some mathematics from assessment tasks, but the results should yield information that can be used to improve students' access to subsequent mathematical knowledge. The results must be timely and clearly communicated to students, teachers, and parents. School time is precious. When students are not informed of their errors and misconceptions, let alone helped to correct them, the assessment may reinforce misunderstandings and waste valuable instructional time. When the line between assessment and instruction is blurred, students can engage in mathematical tasks that not only are meaningful and contribute to learning, but also yield information that the student, the teacher, and perhaps others can use. An oft-stated goal of reform efforts in mathematics education is that visitors to classrooms will be unable to distinguish instructional activities from assessment activities.

Aligning Assessment With Opportunity to Learn

The notion that some students can learn mathematics and others cannot must end; mathematics is not reserved for the talented few, but is required of all to live and work in the Twenty-first century. Assessment should be used to determine what students have learned and what they still need to learn to use mathematics well. It should not be used to filter students out of educational opportunity. Designing assessments to enhance equity will require conscientious rethinking, not just of what we assess and how we do it, but also of how different individuals and groups are affected by assessment design and procedures. The challenge posed by the equity principle is to devise tasks with sufficient flexibility to give students a sense of accomplishment, to challenge the upper reaches of every student's mathematical understanding, and to provide a window on each student's mathematical thinking.

Some design strategies are critical to meeting this challenge, particularly permitting students multiple entry and exit points in assessment tasks and allowing students to respond in ways that reflect different levels of mathematics knowledge or sophistication. There are no guarantees, however, that new assessment will be fairer to every student; that every student will perform better on new assessments; or that differences between ethnic, linguistic, and socioeconomic groups will disappear. Although this is the hope of the educational reform community, it seems clear that this hope must be balanced by a spirit of empiricism. There is much more to be learned about how changes in assessment will affect long standing group differences.

Equity implies that every student must have an opportunity to learn the important mathematics knowledge that is assessed. Obviously, students who have experience reflecting on the mathematics they are learning; presenting and defending their ideas; or organizing, executing, and reporting on a complex piece of work will have an advantage when called upon to do so in an assessment situation. Especially when assessments are used to make high-stakes decisions on matters such as graduation and promotion, the equity principle requires that students be guaranteed certain basic safeguards.

Students cannot be assessed fairly on mathematics content that they have not had an opportunity to learn. Assessments can contribute to students' opportunities to learn important mathematics only if they are based on standards that reflect high expectations for all students. There can be no equity in assessment as long as excellence is not demanded of all. If we want excellence, the level of expectation must be set high enough so that, with effort and good instruction, every student will learn important mathematics (Wilkins & Harnisch, 1997).

The program goals for Project 2061 (AAAS, 1995) focuses on the design of project materials that make sense for all students. This particular project has addressed the equity issue right from the start with its initial publications that envisioned all students will learn the same content, no matter what their background. The project is addressing issues of inclusiveness and diversity goals by commissioning a blueprint on equity to be published in 1997 together with blueprints on other implementation issues.

We have much to learn about how to maintain uniformly high performance standards while allowing for assessment approaches that are tailored to diverse backgrounds. Uniform application of standards to a diverse set of tasks and responses poses an enormous challenge that we do not yet know how to manage fairly and effectively. Nonetheless, the challenge is surely worth accepting.

ASSESSMENT LINKAGES WITH CURRICULUM AND PROFESSIONAL DEVELOPMENT

Six years after its original standards publication, NCTM (1995) released *Assessment Standards for School Mathematics*, a booklet addressing curriculum and evaluation. This document discusses desired attributes of student assessment that includes alignment with the curriculum, multiple sources of information, appropriate assessment methods, and making instructional decisions and evaluating the mathematics program. Detailed standards are provided for assessing student learning from a daily monitoring of student progress to an evaluation of their achievement pattern. Additional performance tasks appropriate to monitor mathematics learning based on an NCTM standards-linked curriculum are given. Suggestions for reporting student achievement to the public are provided with an emphasis that this be done based on descriptive summary reports and suggested work samples.

In the current context of academic standards (i.e., national, state, district, and classroom) reform, assessments are expected to be aligned to the agreed-upon standards of performance. Assessments are to match instructional goals and reflect positive instructional effects. Changes in instructional practice based on current research should be reflected in improved student performances.

Improved assessment is required to round out and support the changes under way in mathematics education, both in the kinds of mathematics taught and in the ways in which they are taught. Given these situations, it is easy to see that assessment is an important partner in the learning network, along with curriculum and professional development of teachers. As assessment is often the focus in determining what students learn and how teachers teach, it must be tied closely with the new vision of teaching and learning.

How do students learn meaningful and important mathematics? Research has shown that students learn important mathematics when they use mathematics in relevant contexts in ways that require them to apply what they know and to extend their thinking. It is very important to remember that students think when they are learning and that they learn when they are thinking. Outstanding educators have long observed that mathematics comes alive for students when it is learned through experiences they find meaningful, connected, and valuable. Children today learn best and most enduringly by engaging mathematics actively, by reflecting on their lab or performance experience, and by communicating with others about it. Learners want to make sense of their setting and world, and mathematics is a wonderful tool to use.

Teacher Collaboration

Teachers value collegiality over direct instruction from a supervisor (Weiss, 1978). Bringing teachers together so they can support each other and provide each other with information is consistent with the current professional development programs. The National Science Foundation (NSF) Elmhurst College mentoring project (Harnisch, 1996) represents an administrative structure devoted to bring teachers together to foster their communication with each other, both formally at institutes and informally between individual teachers and their mentors. These mentoring networks served as the primary means to inform and support math teachers as they worked toward changes reflected in the NCTM standards.

Teachers, when implementing the new professional teaching standards, more often will serve as guides or coaches than as tellers of information. They will be able to focus on helping students pose questions, gather data, and evaluate solutions. Teachers will have access to more information in different formats to suit the needs of individual learners (e.g., CD ROM, videodisk, worldwide databases, and networks of consultants). Instruction will no longer be limited to a single teacher in a single site teaching only with the information that is physically available in that place. For example, there are instances of NASA staff serving as on-line consultants to junior high science students, virtual discussion groups between groups of students from different campuses, and discussions with textbook authors. As the nature of instruction and the way teachers work changes, a changing notion of professional development is needed.

Traditionally, professional development is viewed as a workshop or institute teachers attend to learn methods for implementing curriculum or improving practice. Based on the NSF mentoring project and other school based teacher development efforts, teachers were engaged in a broad range of professional development experiences to learn about, and to become proficient in, implementing change. Communities of teachers, networks, and leadership roles have all contributed to this expanded notion of professional development. The professional development change we are talking about is a move from "quick fixes for Monday morning" to an understanding that significant change will take time and commitment.

The primary reason this view of professional development is needed is that the NCTM effort reflects teachers implementing curriculum based standards focusing on student inquiry, problem solving, reasoning, and applications. This view of instruction is in marked contrast with what teachers have done over the past decade. Another possible reason for this changing notion of professional development is that classroom teachers are being valued as specialists with expertise that others in the educational enterprise do not have. Another very interesting consideration is the integrity that teachers are able to bring to the situation in terms of explaining and describing how they can work with students.

Teachers are deeply interested in doing their best for their students. Math teachers feel a sense of professional responsibility to make sure the foundational ideas shared in an introductory course are understood, thus allowing their students to perform well in an advanced algebra class. For example, teachers hearing a message of how to successfully integrate the use of a graphic calculator with a beginning algebra class find this very meaningful. The dynamics among teachers in math programs are changing. More teachers are finding ways to engage in cooperative planning, experimentation and reflection. Teachers with this professional development experience report their willingness to try new ideas or take some risks.

Teachers are a fundamental key to assessment reform. As evaluation of student achievement moves away from short answer recall of facts and algorithms, teachers will have to become skilled in using and interpreting new forms of assessment. As a result, teachers' professional development at both the preservice and inservice levels will become increasingly important. To the extent that communication is a part of mathematics, differences in communication skill must be seen as differences in mathematical power.

Mathematics students learn important lessons when they work in teams, combining their knowledge and discovering new ways of solving problems. What often occurs is that there is no single right answer, only several possibilities that unfold into new questions. Students need opportunities to advance hypotheses, to construct mathematical models, and to test their inferences by using the mathematics of estimation and uncertainty alongside more traditional techniques of school mathematics. With the use of graphing calculators, students have the opportunity to explore complex, real life problems. Difficulties associated with computational and symbol manipulation need no longer block the development of problem solving or mathematical modeling skills.

Many teachers across this country are reflecting on this new vision of learning and teaching. Summer institutes provide teachers with the opportunities to learn how to integrate these technological tools into their math and science classrooms. For example, summer technology workshops for math teachers are being offered by the author this summer with sponsorship from professional businesses and staff development organizations. Current assessments, however, often do not support this vision and often work against it. For the past several decades, educational assessment in the United States has been driven largely by practical and technical concerns rather than by educational priorities.

A narrow focus on technical criteria, primarily reliability, also works against good assessment. For too long, reliability meant that examinations composed of a small number of complex problems were devalued in favor of tests made up of many short items. Students were asked to perform large numbers of smaller tasks, each eliciting information on one facet of their understanding, rather than to engage in complex problem solving or modeling mathematics work that is most important. In the absence of expressly articulated educational principles to guide assessment, technical and practical criteria have become de facto ruling principles. The content, learning, and equity principles are offered not to challenge the importance of these criteria, but to challenge their dominance and to strike a better balance between educational and measurement concerns. An increased emphasis on validity, with its attention to fidelity between assessments, high quality curriculum and instruction, and consequences, is the tool by which the necessary balance can be achieved.

Assessment Vision

The boldness of our vision for mathematics assessment should not blind us either to the obstacles educators will face or to the limitations on resources we possess for making it come about. Even if new assessments were to magically appear and to be implemented across the nation, many substantial problems will remain. Examples of important, unresolved issues abound.

Open-ended problems are not necessarily better than well-defined tasks. The mere labels "performance assessment" and "open-ended" do not guarantee that a task meets sound educational principles. For example, open-ended problems can be interesting and engaging, but mathematically trivial. Performance tasks can be realistic and mathematically appropriate, but out of harmony with some students' cultural backgrounds.

The equity principle implies that all students are provided an opportunity to learn the mathematics that is assessed and that schools are held to "school delivery standards" to ensure that students are provided with appropriate preparation, particularly for any high stakes assessment. Many would argue, however, that past remedies designed to improve schools often failed precisely because the emphasis was placed on the resources schools should provide, rather than the outcomes that schools should achieve.

The equity principle also requires some consideration of consequences for schools based on the way assessments are used. Fair inferences can be drawn and comparisons can be made only when assessment data include information on the nature of the students served by the school, students' opportunities to learn the mathematics assessed, and the adequacy of resources available to the school. Assessments based only on partial data (typically outcome scores on basic skills) can seriously mislead the public about how schools are performing and how to improve them.

On the job and in the real world, knowledge is frequently constructed and validated in group settings rather than through individual exploration. Mathematics is no exception. Learning and performance are frequently improved in group settings. Hence, assessment of learning must reflect the value of group interaction. The challenge of fairly appraising an individual's contribution to group efforts is immense, posing unresolved problems both for industry and education.

New performance based assessments introduce significant challenges for the mathematical expertise of those who score assessments and for the guidelines used in scoring. Problem solving legitimately may involve some false starts or blind alleys; students, whose work includes such things, are doing important mathematics, and their grades need to communicate this in an appropriate fashion. All graders must be alert to the unconventional, unexpected answer that may contain insights that the assessor had not anticipated. Of course, the greater the chances of unanticipated responses, the greater the mathematical sophistication needed by those grading the tasks.

As assessments become more complex and more connected to real world tasks, there is a greater chance that the underlying assumptions and points of view may not apply equally to all students, particularly when differences in background and instructional histories are involved. Despite good intentions and best efforts to make new assessments fairer to all students than traditional forms of testing, preliminary research does not confirm the corollary expectation that group differences in achievement will diminish. Indeed, recent studies suggest that differences may be magnified when performance assessment tasks are used.

CONCLUSION

Assessments represent an unparalleled tool for communicating the goals and substance of mathematics education reform to various stakeholders. Assessments make the goals for mathematics learning real to students, teachers, parents, policymakers, and the general public, all of whom need to understand clearly where mathematics reform will take America's children and why they should support the effort. Assessments can be enormously helpful in this re-education campaign, especially if the context and rationale for various tasks are explained in ways that the public can understand. The challenges and opportunities that we face are how to improve our assessment practices that are linked with instruction and how to link assessment results to national standards.

Improved assessment can lead to improved instruction. Assessment can play a key role in exemplifying the new types of mathematics learning students must achieve. Assessments can indicate to students not only what they should learn, but also the criteria that will be used in judging their performance. For example, a classroom discussion of an assessment in which students grade some (perhaps fictional) work provides a purely instructional use of an assessment device. The goal is not to teach answers to questions that are likely to arise, but to engage students in thinking about performance expectations.

Assessment can also be a powerful tool for professional development as teachers work together to understand new expectations and synchronize their expectations and grades. Teachers are rich sources of information about their students. With training on methods of scoring new assessments, teachers can become even better judges of student performance.

Meaningful evaluation of educational programs depends to a large degree on the quality of educational indicators used. By knowing the capacity of the indicators to assess the subject and the correlation between various indicators, it is possible to make predictions from one indicator to another. When we have highly correlated indicators, the question of greatest interest is no longer a question of which is best, but which is least costly in time and/or resources.

In school based settings, the assessment of student learning is generally built around the specified content domains addressed in the curricular frameworks and course outlines. In general, teachers design a course of instruction based on these guidelines and use a variety of methods for their students to experience the subject matter. Having a high correlation between what is taught and the formal curriculum makes large scale assessment possible. Indicators are developed that measure student learning along the articulated curriculum lines as there is typically a high correlation between instruction and the stated curriculum. Because of the typical high relationship between instruction and stated curriculum, inferences can be drawn about the effectiveness of instructional strategies in school systems, schools, and classrooms along curricular lines.

In the School Design Collaborative, Harnisch and Jen (1997) have been developing a set of reporting formats that allow teachers and administrators to review test performances on multiple indicators. The process used to determine whether the students have achieved the goals set for them, given the indicators employed, range from the results of simple observations to group-administered standardized tests and from a short homework assignment to the performance and analysis of a complex problem. Test scores, portfolios, and performances are all indicators of learning. Selecting the indicators best suited to specific purposes is at the core of the assessment debate. In our discussion with professional support staff for the School Design Collaborative, the issue of determining the appropriate indicators for measuring the effectiveness of program outcomes was debated. Sponsors had their preferences for objective tests while the consultants tended to prefer their criterion-based assessments with examples of student work.

Assessment should be a tool for educational improvement, providing information that allows educators to determine which practices result in desired outcomes and which do not. With a focus on outcomes, the teachers and schools are free to use whatever methods prove practical in achieving student academic progress. The assumption is that effective teaching, whatever form it takes, will lead to student gains. In other words, the methods used by teachers to produce greater student achievement can be identified and shared. This reporting system is an entirely new approach to using normed data. Our efforts focus on providing information on educational effects that measures of ability cannot. High achievement scores do not necessarily mean student progress, but high changes in student achievement do. When the focus is on the gains that children make from year to year, regardless of where they start, the quality of the schools can be described in terms of their ability to provide opportunities for all students—the advanced learners as well as the slower learners. This approach is often described as the valued-added approach. As Astin (1982) states, "the basic argument underlying the value-added approach is that true excellence resides in the ability of the school or college to affect its students favorably, to enhance their intellectual development, and to make a positive difference in their lives" (p. 14).

Improved assessment is not a panacea for the problems in mathematics education. Our findings neither diminish nor reject important, time honored measurement criteria for evaluating assessment, nor do they suggest that changes in assessment alone will bring about education reform. Clearly, they will not (Brandt, 1992; Harnisch & Mabry, 1993; Linn, Baker & Dunbar, 1991).

We can say with assurance that if old assessments remain in use, new curriculum and teaching methods will have little impact. Moreover, if new assessments are used as inappropriately as some old assessments, little good will come of changes in assessment. The issue is not whether one form of assessment is intrinsically better than another. No assessment model is suited for every purpose. The focus of our effort should be about choosing appropriate indicator variables and applying the most suitable model to interpret them.

It will take courage and vision to stay the course. As changes in curriculum and assessment begin to infiltrate the many jurisdictions of the U.S. educational system, these changes will, at the outset, increase the likelihood of mismatches among the key components of education: curriculum, teaching, and assessment. It is not unlikely that performance will decline initially if assessment reform is not tightly aligned with reform in curriculum and teaching.

The idea of linking assessment to academic standards is on a fast track. In the spring of 1997, three new statewide assessment systems have been introduced (Florida, Utah, and Kansas) to make this linkage. President Clinton has made national testing, linked to national standards, one of the elements in his administration's plan to improve education. The Department of Education is busy creating national tests linked to national standards in reading for fourth graders and mathematics for eighth graders. The 90-minute assessments, modeled on the National Assessment of Educational Progress, are due in the classrooms by 1999. The idea of national tests will rekindle a hornet's nest of controversy as to who controls the educational agenda. What we need is a coherent vision for our subject matter areas, rather than a national curricular road map (Shavelson, 1997).

Given the national testing movement, we begin to see how mathematics education is entering a period of transition in which there will be considerable exploration. Inevitably both successes and failures will occur. No one can determine in advance the full shape of the emerging national assessments. Mathematics education is, in this respect, an experimental science, in which careful observers learn as much from failure (and from the unexpected) as from anticipated success. The necessary change will be neither swift nor straightforward. Nevertheless, we cannot afford to wait until all questions are resolved. It is time to determine what information is sufficient to each purpose before deciding upon the form of assessment to be used. It is time to put educational principles at the forefront of mathematics assessment.

The development of new curriculum standards and assessments should not be limited to experts. Discussion on what knowledge is most important for all students to learn and how they should be tested requires open, public deliberation (McDonnell 1997). The design and development of curriculum standards requires a real world perspective where different points of view can be shared in shaping the statements. A long term commitment, politically and educationally, is needed that recognizes that implementing a reform as complex as new curriculum and testing will take time.

REFERENCES

American Association for the Advancement of Science (AAAS). (1989). *Science for all Americans: A Project 2061 report on literacy goals in science, mathematics, and technology.* Washington, DC: Author.

American Association for the Advancement of Science (AAAS). (1993). *Benchmarks for science literacy.* New York: Oxford University Press.

American Association for the Advancement of Science (AAAS). (1995). *Project 2061: Science literacy for a changing future, a decade of reform.* Washington, DC: Author.

Astin, A. W. (1982). *Excellence and equity in American education.* Washington, DC: National Commission on Excellence in Education.

Blank, R., & Pechman, E. M. (1995). *State curriculum frameworks in mathematics and science: How are they changing across the states?* Washington, DC: Council of Chief State School Officers.

Brandt, R. (1992). On performance assessment: A conversation with Grant Wiggins. *Educational Leadership, 49* (8), 35-37.

Darling-Hammond, L., & Wise, A. E. (1985). Beyond standardization: State standards and school improvement. *Elementary School Journal, 85* (3), 315-336.

Harnisch, D. (1994). Performance assessment in review: New directions for assessing student understanding. *International Journal of Educational Research, 21*, 341-350.

Harnisch, D. L. (1995). *A framework for reporting performance on problem-solving assessments: SPP charts and classroom applications.* Proceedings from the 1995 annual meeting of the Society for International Technology Education(SITE). Available on SITE produced CD and from the SITE web site.

Harnisch, D. L. (1996). *A mentoring network for increasing technology-based mathematics instruction: Year 3 report.* University of Illinois, Office of Educational Testing, Research, and Service. Report prepared for National Science Foundation sponsored program at Elmhurst College, Elmhurst, IL.

Harnisch, D. L. , & Mabry, L. (1993). Issues in the development and evaluation of alternative assessments. *Journal of Curriculum Studies, 25* (2), 179-187.

Harnisch, D. L. , & Switzer, D. M. (1991). Teachers' perceptions of the instructional uses of tests. In R. E. Stake (Ed.), *Advances in Program Evaluation: Effects of Mandated Assessment on Teaching* (Vol. 1, Part B, pp. 163-188). Greenwich, CT: JAI Press.

Harnisch, D. L., & Jen, C. Y. (1997). *Profiles of Student Performance: St. Elizabeth Elementary School. School Design Collaborative Report.* Urbana, IL: University of Illinois.

Illinois Goals Assessment Program (IGAP). (1995). *Technical summary of score reports.* Springfield, IL: Illinois State Board of Education.

Linn, R. L. (1983). Testing and instruction: Links and distinctions. *Journal of Educational Measurement, 20*, 179-189.

Linn, R. L., Baker, E. L., & Dunbar, S. (1991). Complex, performance-based assessment: Expectations and validation criteria. *Educational Researcher, 20* (8), 15-21.

Madaus, G. F. (1985). Public policy and the professionDYou've never had it so good? *Educational Measurement and Practice, 4* (4), 5-11.

Marzano, R. J. & Kendall, J. S. (1995). *The systematic identification and articulation of content standards and benchmarks.* Aurora, CO: Mid-continent Regional Education Laboratory.

Mathematical Sciences Education Board. (1989). *Everybody counts.* Washington, DC: National Academy Press.

Mathematical Sciences Education Board. (1993). *Measuring what counts: A conceptual guide for mathematics assessment.* Washington, DC: National Academy Press.

McDonnell, L. M. (1997). *The politics of state testing: Implementing new student assessments.* [Monograph]. Los Angeles: CRESST.

McLeod, D. B., Stake, R. E., Schappelle, B. P., Mellissionos, M., & Gierl, M. J. (1996). Setting the standards: NCTM's role in the reform of mathematics education. In S. A. Raizen & E. D. Britton (Eds.), *Bold ventures: Case studies of U. S. innovations in mathematics education* (Vol. 3, pp. 13-132). Boston: Kluwer Academic Press.

National Academy of Science. (1982). *Science and mathematics in the schools: Report of a convocation.* Washington, DC: National Academy Press.

National Commission on Excellence in Education. (1983). *A nation at risk: The imperative for educational reform.* Washington, DC: U.S. Government Printing Office.

National Council of Teachers of Mathematics (NCTM). (1989). *Curriculum and evaluation standards for school mathematics.* Reston, VA: Author.

National Council of Teachers of Mathematics (NCTM). (1991). *Professional standards for teaching mathematics*. Reston, VA: Author.

National Council of Teachers of Mathematics (NCTM). (1995). *Assessment standards for school mathematics*. Reston, VA: Author.

National Research Council (NRC). 1996. *National science education standards*. Washington, DC: Author.

National Science Board, Commission on Pre-college Education in Mathematics, Science and Technology. (1983). *Educating Americans for the 21st century*. Washington, DC: National Science Foundation.

O'Day, J. A., & Smith, M. S. (1993). Systemic reform and educational opportunity. In S. Fuhrman (Ed.), *Designing coherent education policy: Improving the system* (pp. 250-312). San Francisco: Jossey-Bass.

Ravitch, D. (1995). *National standards in American education: A citizen's guide*. Washington, DC: The Brookings Institution.

Shavelson, R. (1997, May 7). The splintered curriculum. *Education Week*, 38.

Shepard, L. (1991). Interview on assessment issues with Lorrie Shepard. *Educational Researcher, 20* (2), 21-23.

Stake, R. E. (1995). The invalidity of standardized testing for measuring mathematics achievement. In T. A. Rombery (Ed.), *Reform in school mathematics and authentic assessment* (pp. 173-235). Albany: State University of New York Press.

Stake, R. E., & Raizen, S. A. (1997). Underplayed issues. In S. A. Raizen & E. D. Britton (Eds.), *Bold ventures: Case studies of U.S. innovations in mathematics education* (Vol. 1, pp. 131-153). Boston: Kluwer Academic Press.

Weiss, I. R. (1978). *Report of the 1977 national survey of science, mathematics and social studies education*. Research Triangle Park, NC: Research Triangle Institute.

Wilkins, J. L., & Harnisch, D. L. (1997, March). *Item response patterns and classroom indicators of opportunity-to-learn*. Paper presented at the Annual Meeting of the American Educational Research Association, Chicago.

Wilson, B., & Corbett, D. (1991). Two state minimum competency testing programs and their effects on curriculum and instruction. In R. E. Stake (Ed), *Advances in Program Evaluation* (Vol. 1, Part B, pp. 7-40). Greenwich, CT: JAI Press.

PART III

INCREASED CHOICE IN PUBLIC EDUCATION

SCHOOL CHOICE AND STUDENT PERFORMANCE IN PUBLIC SCHOOLS

John F. Witte

INTRODUCTION

Educational choice has been a "hot button" issue for over a decade in the United States. During that period, the battle over choice has shifted. Initially, as a number of states began exploring choice options in the middle 1980s, opposition centered on public school choice proposals, such as inter-district choice and open enrollment. During the last decade, public school choice options seem to have become accepted by almost all, including leading politicians in both parties, most educational special interest groups, and both national teachers' associations. A very intense battle remains, however, over private school choice. Of all the potential educational issues on which Presidential candidates might differ, vouchers for students to attend private schools was one of the only issues that drew a clear distinction between Bill Clinton and Bob Dole in the 1996 campaign.

Before this period of intense scrutiny, many public school districts had experimented with choice options within their districts and between districts. This paper focuses on those public school choice options and what we know

Advances in Educational Administration, Volume 5, pages 167-189.

ISBN: 0-7623-0081-7

and do not know about their effects on student performance. The topic is extremely important because, despite all the media attention to private school choice (especially vouchers), public school choice programs affect hundreds, if not thousands of times more students than private school choice programs. In a sense, this paper backs up in time, asking what we know about the effects of public school choice somewhat after the fact.

PUBLIC SCHOOL CHOICE PROGRAMS

Public school choice can be defined broadly as any arrangement that deviates in assigning pupils to schools from one based exclusively on geographic location of students and that provides for some degree of family choice in school assignment. For most of our history, school assignment was a trivial issue. As public schools developed as professional organizations in the latter half of the nineteenth century, the only real assignment issue was the divergence of private schools, especially Catholic schools. Following the third Plenary of American Bishops in 1883, Catholic parishes were directed to build schools and Catholics were to attend those schools rather than the Protestant dominated public schools.[1] The vast majority of students, however, attended local public schools, most in single room buildings, with all ages mixed together. One of the major thrusts of the professional public school movement was to create consistently graded classrooms and curriculum. Where enough students existed, students were separated into distinctive rooms based on grades. With the exception of a small private school system, which school to attend in one's village, town, or neighborhood was not an issue.

The earliest examples of out of attendance area schools were experimental, specialized, and clearly elite schools. They would today be termed either *magnet* or *charter* schools. Examples date to the nineteenth century with the creation of the Boston Latin School and a series of four schools for gifted and talented students in New York City: Stuyvesant, Brooklyn Technical School, Bronx School of Science, and Laguardia High School for the Arts (of *Fame* lore). Perhaps the original charter school was the University of Chicago Lab School, created in concert with the philosophy of John Dewey as a campus school to provide a training environment for teachers and as an experimental school for education scholars.

Although these schools were important in their local districts and have long histories of success and successful graduates, they remained a very minor anomaly in comparison to the growing public school system. Of more consequence was the movement to establish teacher accreditation, create uniform and professional curricula, and consolidate school districts, jettisoning the traditional one room school, which usually served as its own district. Assignment and community support for all but a handful of students was based on where students lived. That meant the neighborhood elementary school and the closest middle and high school.

Although there are no precise statistics apart from statistics that will be presented next based on national surveys of middle and high schools, traditional geographic assignment probably still exists for the majority of public school students in America. Survey data indicate that the majority of students in all districts, and clearly those residing in middle sized cities, suburbs, towns, and rural areas still attend schools based on residential attendance area.

For many students in large metropolitan areas, however, attendance boundaries defined by neighborhoods began to breakdown as a result of the effort to desegregate schools following *Brown v. Board of Education* (1954). Cities of many sizes were affected in the South, while primarily large school districts were affected in the North. School choice was touted as a mechanism to achieve integration without resorting to compulsory busing. Magnet schools were the principal vehicle. As Rossell (1990) has carefully described, they provided a carrot rather than the stick of compulsory busing. The idea was to create special schools, that would attract families of all races who would voluntarily have their children attend these schools. Specifically, the hope was to circumvent white and middle class flight to the suburbs in the wake of court ordered desegregation plans.

Even the most ardent advocates of integration today admit success was dubious (Orfield & Eaton, 1996). The growth of magnet schools in large districts, however, was spectacular. Although it began in the 1970s, spurred on by desegregation orders (or the threat of such orders), in the 1980s the movement seemed to take on a life of its own, providing choice for its own sake to increasingly large minority student populations (Blank, 1990). Magnet schools and magnet school programs remain today as the primary mechanism of school choice in America, public or private.

Another choice offshoot of the integration movement was interdistrict choice programs. A key battle and turning point in the integration movement came when desegregation advocates, acknowledging the problems created by white flight, attempted to expand the scope of integration orders to include multiple districts, specifically suburban districts surrounding large, increasingly minority city districts. The battle was more or less lost in Detroit (*Milliken v. Bradley*, 1974). Enough uncertainty and belief in integration goals persisted to allow for the creation of a number of voluntary interdistrict transfer programs designed to increase integration on a metropolitan basis. Such programs were created in St. Louis, Kansas City, Louisville, and Milwaukee. They allowed city, non-white students to attend suburban schools, with white students transferring from the suburbs to the city.

During the same period in the 1980s, a number of other pioneering public school choice programs were being proposed. Minnesota was the most prominent state. During the late 1980s and early 1990s, Minnesota passed legislation creating state wide alternative learning centers (1987), open enrollment (1986), and then charter schools (1991). They had also been the first state to enact tuition tax credits for private schools in 1967.[2] In 1985 they enacted a post-secondary options program that allowed high school students to attend public or private colleges while in high school, receiving college credit and having the state pay the college tuition.

All three of the Minnesota public school choice programs have been emulated by a number of districts and states. Most state legislation in the last decade, however, has focused on open enrollment and charter schools.[3] Scholars and various information dissemination organizations have tracked the expansion of open enrollment and charter school programs. The number and types of laws and programs are presently being charted. Unfortunately, knowledge of program developments is not the same as knowledge of their effects.

The four major types of public school choice to be discussed in this paper are presented in Table 1. The primary purpose of the paper is to assess what is known about the effects of these programs on student achievement, broadly defined. Achievement in choice schools, however, cannot be considered apart from the question of who chooses? Which students are most likely to enroll in non-attendance area schools is important as an independent policy issue, but it also must be factored into accurate estimates of achievement effects of choice verses non-choice schools. One of the major considerations in determining if we know the achievement effects of choice is whether we can adequately measure the potential selection bias that might result (Witte, 1992).

Table 1 Major Public School Choice Programs

Public School Choice Options	*When Initiated?*	*Where?*	*Motivation?*
(1) Magnet and "Choice" Schools	19th Century; Expanded 1970s Expanded 1980s to date	Boston/NY; Large districts Large, medium districts	Specialized and elite schools; Integration; Choice, innovation
(2) Inter-District Choice	1980s	Metropolitan areas: St. Louis, Kansas City, Louisville, Milwaukee	Voluntary metropolitan integration
(3) Open Enrollment	1980s-1990s	Minnesota initially; 1996= over 20 states	Choice
(4) Charter Schools	1990s	Minnesota initially; 1996= 25 states	Innovation, choice, autonomy

The analysis will proceed by choice category, but the studies discussed do not always utilize the categories I have described. In addition, although I do not independently describe the effects of alternative choice schools, districts often have established such schools, either as magnet schools or as simply choice schools. Thus they are included in national studies that inquire about choice schools in districts.

The general theses of this paper are that we know a considerable amount about the expansion of these alternatives, but we know less about the effects of these programs on student achievement. What we know, however, is more positive than negative. The paper ends with a question and a call. Why is it we know less, and experts seem less concerned, about public than private school choice? The question is especially relevant because unlike private school choice, public school choice options seem to be widely accepted and continue to expand. This suggests that a major effort is necessary to evaluate current public school choice programs.

MAGNET AND OTHER CHOICE SCHOOLS

Although most programs and studies use the words *magnet schools* to mean schools developed with special programs that are designed to attract out of attendance area students, some studies and districts simply refer to such schools as *choice schools*. The terms will be used interchangeably from this point forward. This section considers and is organized around a number of studies differentiated by the data employed. It begins with a description of the expansion of magnet programs, then it describes evidence based on national studies, followed by evidence based on data from Milwaukee. Several other districts are mentioned in conclusion, although only to make the point that no useful performance data exist in these and many other districts with magnet schools.

The Expansion of Magnet and Choice Schools

Public school choice exploded in the 1980s and 1990s. It began with the expansion of magnet schools and programs, followed by interdistrict choice programs and open enrollment, and has recently included quasi-public charter schools. Early studies of magnet schools focused almost exclusively on their initial purpose, voluntary integration. One of the most thorough and systematic studies of magnet schools as a tool for integration carefully evaluated their effectiveness in terms of furthering integration, curriculum differences, costs, and equity issues. Lacking systematic data, however, there was no effort made to assess the effects on student performance (Rossell, 1990).

There are numerous anecdotal stories concerning the effects of choice schools on performance. They are usually success stories, such as Fliegel's *Miracle in*

East Harlem (1993), or the magnet program in Montclair, New Jersey. These successes, however, were almost never carefully analyzed using prior achievement, adequate controls, or any type of comparison group.[4]

Similarly, Blank (1990) has written an excellent study of the expansion, location, and types of magnet schools in the 1980s. He clearly documents the dramatic expansion of magnet schools and magnet school programs beyond their initial use as an aid to integration. The study was based on detailed data collected in 1982-1983 and 1988-1989 from 15 school districts employing magnet schools. The growth in the number of magnet schools in these districts over the six year period was considerable. In 1982-1983 there were 148 elementary and 95 middle and high school magnets in these districts. By 1988-1989 the number of elementary magnet schools had increased to 244 and the number of middle and high school magnets increased to 147 (Table 2.1, p. 82).

Blank's study, however, produced only very limited information on the effects of magnets on school performance. Although student outcome measures were collected, Blank concluded: "...it was determined that no statistical comparisons of outcomes across the studies, or comparisons with a national standard would be possible" (p. 93). Although an effort was made to compare outcome measures across programs, only one district had over time measures of achievement, controls for student characteristics, and a control group. Only three districts had over time measures and any controls on student characteristics. The single district with adequate data and controls was Montgomery County in Maryland. The study of elementary school magnets found positive increases in reading and math achievement for those students who had been in the program the longest (p. 95).

Studies Based on National Level Data Bases

Several recent studies based on the National Educational Longitudinal Study (NELS) data do provide some information on who attends and the achievement effects of public magnet and choice schools. NELS is a national study that began in 1988 with a random sample of over 26,000 eighth grade students in over 1,000 schools nationwide. It included both public and private school students, interviews with parents, teachers, and administrators, and standardized tests designed for the study that were given to all participating students. The students were followed into tenth grade in 1990, twelfth grade in 1992, and then into post-secondary school in 1994. The only currently published studies are based on the eighth to tenth grade transition and the changes in achievement between these periods. The three published studies on public choice schools using NELS provide evidence on who attends public choice schools, the effect on achievement as measured by standardized test results, and some limited evidence on the attitudes of students toward education.

A study headed by the late James Coleman (Plank, Schiller, Schneider & Coleman, 1993) analyzed both the attendance in magnet and choice schools and the achievement effects of choice students in the NELS study. Their study supports a finding, which is confirmed by other studies, that African American, Hispanic, and lower socioeconomic status (SES) students tend to attend public choice schools (Asian students are more highly represented in private schools). That result, however, is clearly a byproduct of the location of most choice schools. They split public schools into: assigned, choice (schools allowing student selection or schools with admissions tests), magnet, and vocational. In urban areas, according to Plank et al. (1993, p. 119), the percentage of students in assigned schools was 45.7 percent with 12.8 percent in choice schools, 15.0 percent in magnet schools, 6.4 percent in vocational schools, and 20.1 percent in private schools. In suburban areas, the percentage of students in assigned schools was 72.2 percent, with 10.9 percent in choice schools, 3.4 percent in magnet schools, 1.9 percent in vocational schools and 2.4 percent in private schools.

Focusing on the differences between urban and suburban schools, which account for the majority of students, it is obvious that school choice in general is more likely to occur in urban areas and that magnet schools are predominantly an urban phenomenon. This makes it difficult to disentangle the national overrepresentation of Blacks, Hispanics and lower-SES students in choice schools. Both of the other studies report similar results concerning who attends choice schools.

Plank et al. (1993) also present some information on magnet and choice schools on prior levels of achievement and educational aspirations. The results are not very clearly presented and are difficult to interpret because they present profiles of each school type rather than population percentiles. It is clear, however, that although private schools enrolled students with much higher prior grades, test scores, and aspirations, public school choice students were a mixed lot. For example, while choice students were similar to assigned school students, magnet students seemed to have lower test scores (more students in the bottom quartile) and more students with only high school level aspirations. In contrast, they had better grades, however, than students attending assigned, comprehensive schools. No matter how one looks at these numbers, however, they do counter the conclusion of a widely cited, unpublished study that made a very strong argument in 1988 that magnet schools were creaming off the best students (Moore & Davenport, 1988).

Plank et al. (1993) also reported on the effects of choice schools on achievement on tenth grade test scores controlling for eighth grade scores, race, family background characteristics, student educational aspirations, and school location. The type of choice school is then included as a series of dummy variables to distinguish it from assigned schools. The authors summarize their results in terms of rank order effects (for simplicity for non-technical readers).

The direct regression effects, however, are more telling. First, the signs of the effects for magnet and choice schools switched from positive, for a composite of four tests, to negative coefficients for mathematics. The only coefficient that was even close to significant, however, was the negative coefficient for choice schools on mathematics (*b*=-.305; *se*=.171). The effects for vocational technical specialty schools versus assigned schools were negative and significant for both the composite and mathematics tests (Plank et al., 1993, Appendix B). Unfortunately, the authors failed to discuss significant levels of coefficients or include the standard deviations of the dependent variables in the article. Thus, assessment of the size of these effects is not possible. In addition, there were no controls for within school clustering of similar students or for selectivity bias.

Another NELS study, using a different school selection design, analyzed only the characteristics of students enrolling in choice verses assignment area schools. Driscoll (1993), attempting to control for the excessive number of choice schools in urban areas, created a matched school design of choice and attendance area schools based on urban, suburban, or rural location. Her study was based on 66 assignment area schools matched with choice schools, which were defined as "... schools with what we believe to be the central feature of choice schools (i.e., selection of schools by students and of students by schools)" (p. 149). Her interest was the characteristics of families selecting choice options.

In terms of eighth grade achievement tests, the Driscoll (1993) study, utilizing analysis of variance concluded that:

> ...there was no real difference in achievement levels between these two sets of schools. Students in choice schools did not do better than those in non-choice schools; in fact, the one small statistically significant difference showed that students in the matched [assignment] schools actually performed better than students in choice schools on the mathematics achievement test (p. 155).

The difference in math scores was actually quite significant, with a mean difference of 17 points, which is significant at the .02 level. This means that, if anything, choice students were doing less well than non-choice students prior to entering choice schools.

On the other hand, parental attitudes for those students picking choice high schools were significantly more positive. These included questions on overall satisfaction, student enjoyment of the school, how much the student was challenged, the worth of homework assignments, and school safety (Driscoll, 1993, p. 157, Table 3). In addition, parents in choice schools responded that they talked more often about the educational plans for their eighth graders and had less contact with prior schools concerning behavioral problems than their assignment school counterparts (p. 159, Table 4). Finally, students reported greater satisfaction and better relationships with teachers than non-choice students (p. 161, Table 5). Thus one could argue that while choice public schools do not attract, or "cream off" the best students in terms of test scores, they do seem to enroll students from families that value and are more involved in education.[6]

Gamoran (1996) has published the most sophisticated study to date using NELS data. It incorporates both public and private school choice. To control for the differential use of choice schools by locale, however, he limited his study to urban schools. To support his decision, he demonstrated, consistent with other NELS studies, that central city students were more likely to attend magnet schools (3 to 1 or more) and private schools (2 to 1 or more). He also reaffirmed 1993 studies of Plank et al. and Driscoll in that African Americans, Hispanics, and those from lower-SES homes were more likely to attend magnet schools than other students.[7] He estimated three types of models and incorporated a wide range of variables defining student and school characteristics. Within school effects of this study will be discussed. The mean effects of being in either Catholic, nonreligious private, or magnet schools were estimated, using ordinary least squares (OLS) regressions; hierarchical linear models; and a two stage instrumental variables model to correct for potential selectivity bias.

The results for magnet schools were striking and very positive for public school choice. In terms of raw change scores, tenth grade magnet school students, despite having greater economic disadvantages, scored the same as comprehensive high school students in math and science, but scored significantly higher in reading and social studies. When these scores were adjusted for prior achievement on eighth grade tests and student characteristics (sex, race, and family composition), the magnet school effects were significant on all tests. Further, the size of the coefficients were only slightly affected when additional variables controlling for school context (racial composition, free lunch percent, and percent from single parent families in the school) were included (Gamoran, 1996, p. 10, Table 4, panels 1 and 2).

Hierarchical linear modeling estimates allow for more accurate estimates of school level effects which may be affected by the clustering of students in schools.[8] Accounting for within school variance using an HLM reduced the effects for math and science to the point they were not significant, but the positive effects of being in a magnet school for reading and social studies remained (Gamoran, 1996, p. 12, Table 5).

When he corrected for selection bias using a set of instrumental variables (Catholic parent, geographic region, and student enrollment in preschool), both the math and social studies coefficients were higher than the comparably modeled single equation OLS results. Thus the selectivity effects were actually negative, which is a function of adverse selection into magnet schools described earlier (Gamoran, 1996, p. 13, Table 6).

The Gamoran study indicates a robust achievement gain effect for public magnet schools. That finding is partially supported by a somewhat less adequate study of choice schools available in a national data base of elementary schools called the Prospects Study.

The Prospects Study was authorized by Congress with the purpose of evaluating Chapter 1 of the Primary and Secondary Education Act. It was designed to follow students in first, third, and seventh grade for five years. By overlapping these grades, a comprehensive study of Chapter 1 effects from kindergarten through grade 12 was hoped for. Unfortunately, the study was defunded by Congress after three years. The data which were collected, however, are valuable in that they are the only grade school level national data base with the type of rich, multi-source data as in NELS or High School and Beyond.

The sample design was schools, with an over sampling of schools with large Chapter 1 populations. Weights are provided that allow one to estimate nationally representative statistics. Within schools, grades (1, 3 ,or 7) were either taken inclusively or sampled depending on the size of the school. A version of the Comprehensive Test of Basic Skills was administered to each student in successive years. Also, student background, family characteristics, attitudes and behavior, and school level data were collected for each child. Because the tests employed in the first grade were not adequate for our purposes here and because the seventh grade sample was small and attrition from the sample was high between grades 7 and 8, we present results for only the third grade cohort. Our focus will be the same as in the earlier studies: What effect does enrollment in magnet or choice schools seem to have on achievement?

The study coded schools as: traditional, non-choice, non-comprehensive choice, comprehensive choice, or magnet schools. Because there were only a handful of magnet schools, and they could overlap with the other choice designations (depending on how the principal answered the query), we have limited our school comparison to non-comprehensive and comprehensive choice verses traditional schools. The results of estimating the fourth grade scores, using OLS methods, and including prior (third grade) tests, and a large set of other independent variables are depicted in Table 2.

The results are mostly as expected. Prior tests were highly significant predictors of subsequent tests. Girls did significantly better on reading, but not on math. African Americans and Hispanics did worse than whites on reading, but not math. Asian students did not differ from whites on reading, but were already by the fourth grade more than 20 percent of a standard deviation ahead of white students on math (after controlling for prior achievement and other variables). Interestingly, when utilizing a value-added model for achievement, disability did not affect achievement changes in either reading or math. Family income and parent education have the expected effects, although multicollinearity increased standard errors and thus only one variable was significant in either equation. In terms of family attitudes and

Table 2. Prospects Study-Effects of Choice Schools on Achievement (Unweighted)

	Reading	*Math*
Prior Tests		
Prior Reading	0.60***	0.16***
Prior Math	0.20***	0.68***
Student SES:		
Gender (1= Female)	2.08***	0.15
African American	-2.20**	-0.94
Asian American	-0.53	4.33***
Hispanic American	-2.52**	-1.03
Other American	-2.27	-0.59
Disabled	1.01	-0.23
Family SES:		
Income	0.30*	0.05
Respondent Education	0.26	0.45**
Respondent Employment	-0.23	-0.33
Respondent Marital Status	-0.34	0.16
Family Attitudes/Behavior:		
Expectations	0.75***	0.69***
School Dissatisfaction	-0.03	-0.05
Parental Involvement—Home	0.09	0.19**
Parental Involvement—Attendance	0.12	0.11
Parental Involvement—Organizations	0.25	0.57***
School Contacts	-0.43***	-.042**
School Type:		
Non-Comprehensive Choice	-0.27	-0.53
Comprehensive—Choice	-1.47	2.26*
Constants	2.01***	-0.43
R^2	0.67	0.63
SE	11.34	11.33
F	232.85***	177.63***
(*df*)	2293,20	2293,20***

Notes: ***probability that $B = 0 < .001$
**probability that $B = 0 < .01$
*probability that $B = 0 < .05$

Source: Author estimates, Prospects Study database.

behavior, expectations were highly predictive of achievement success, as was parental involvement at home and in school in predicting math scores. The negative coefficient in each equation for school contacts probably resulted from negative school contacts, such as those that would occur if a student were having behavioral problems.

Finally, the effects of being in a choice school were somewhat unclear, although the only significant result for mathematics learning was positive for students enrolled in comprehensive choice schools. We have no explanation for the negative sign on reading, which did not approach significance, but still clouds the picture a bit.[9]

In summary, national studies indicate quite clearly that choice options occur primarily in urban areas, where they appear to account for a relatively large and growing number of schools. In addition, because of that fact, in comparison to all schools, students attending magnet and choice schools are more likely to be Black or Hispanic and come from lower-SES homes. One study also suggested that the students had lower prior levels of achievement than students not attending choice schools. That same study, however, also indicated that on a number of important family attitude and behavior measures, the students selecting choice high schools came from educationally more positive homes. The effects on achievement are not fully consistent, but appear to be at least partially positive for both elementary and high school choice schools. The NELS data produced more robust effects at the high school level than the Prospects Study of fourth grade students.

Mixed Findings on Milwaukee Choice Schools

Milwaukee, following a desegregation court order in 1976, was one of the first school districts in the nation to engage in widespread use of magnet schools and magnet programs to aid voluntary integration of its schools. The initial set of magnet schools, generated at all three school levels, tended to be elite magnet schools with an emphasis on gifted and talented programs and high schools either for college bound students or students excelling in the arts. Over time, there was a public backlash by both parents and school staff which led to more egalitarian admission policies for city wide magnet schools and the expansion of magnet schools and programs to a very wide range of schools.[10]

There were two early studies with implications for the impact of magnet schools on performance. In a paper on effective schools and student performance, Witte and Walsh (1990) included variables controlling for specialty school status in Milwaukee. The study was based on 1985 aggregated school data from the Milwaukee Public School system (MPS) and schools in Milwaukee suburban school districts. The effects of being in a magnet school, after controlling for the location of the school (city or suburbs), and the percentage of students in the school on free lunch are given in Table 3.

Table 3. Effects of Milwaukee Specialty Schools on Reading and Math Iowa Test Scores (in Standard Deviations)

	High School	*Middle School*	*Elementary School*
Reading	1.23	0.91	0.52
Math	1.37	0.75	0.81

Source: Witte and Walsh (1990, Tables 2, 3, 5, 6, 7, and 8).

All of the effects indicated higher test scores for students in magnet schools and all the results were highly significant. These effects were somewhat diminished by controlling for parental involvement, effective schools environments, and the extent of control teachers said they had in their schools. The only significant effect of inclusion of these variables on the results for magnet schools was for parental involvement in middle schools. Although the magnet school effects remained positive (.34 standard deviation for reading; .50 standard deviation for math), neither coefficient was significant once parental involvement was included in the equation.

The magnet school effects were considerably larger for high schools. Those schools had average aggregate tests scores more than one standard deviation above non-magnet Milwaukee high schools. Drop out rates reflected a similar very positive effect of being in a magnet school. The study estimated the dropout rate for traditional MPS high schools at 9.6 percent, but only 2.5 percent for magnet schools.

There are several limitations and caveats in this study. First, aggregation of variables at the school level eliminates within school variation that is much higher than between school variance. This inflates the effects as compared to what would be observed for individual students. Second, although we were able to include a proxy for socioeconomic status in the free lunch variable, this is clearly not an adequate test for selectivity, which was undoubtedly occurring on entry into magnet schools.

The second study using these data (Archbald, 1988) highlighted that problem and argued that the magnet system essentially created a two track system in Milwaukee. Because of these problems, while no one contests that magnet schools outperform other MPS schools, the impact of the schools irrespective of selection bias is uncertain.

New data collected as part of an evaluation of the Milwaukee Parental Choice Program from 1991-1994, yields somewhat different results. These data are based on individual students at the elementary and middle school levels. The results are given in Table 4. The table includes two models for reading and math respectively. Model I is based only on student record data base information; Model II also includes variables taken from a survey of MPS parents in 1991. Because not all parents completed the mail survey, sample sizes are considerably

Table 4. Achievement Test Results: Milwaukee Public Schools (Including Heckman Correction)

	Reading		*Math*	
	Model I	*Model II*	*Model I*	*Model II*
Prior Reading NCE	0.510***	0.481***	0.172***	0.155***
Prior Math NCE	0.177***	0.182***	0.559***	0.576***
African American	-3.034***	5.961***	-2.674***	-5.087***
Hispanic	-0.300	-5.980***	1.081	-0.585
Other Minority	-1.968	-4.375*	1.061	-3.652
Gender (1= Female)	1.565***	2.046	0.284	1.808
Grade	-0.536***	-0.015	-1.005***	-1.520***
Free Lunch (1= Yes)	-2.272***	—	-1.506*	—
Parent Income	—	0.170***	—	0.012
Parent Education	—	0.194	—	0.146
Parent Expectations	—	1.291**	—	0.383
MPS Schools:				
pat	0.259	-1.614	0.864	-1.004
cwp	1.288	0.996	0.982	0.397
cw	1.218	1.153	-0.472	-0.048
p-5	0.356	0.726	0.994	2.193
Constant	14.584***	16.644***	6.365***	17.170***
Heckman Correction				
African American	0.285***	0.193**	0.287***	0.171*
Hispanic	0.456***	0.455***	0.429***	0.420***
Other Minority	0.161*	0.064	0.134	0.069
Gender	0.009	0.045	0.028	0.030
Grade	-0.077***	-0.059***	-0.089***	-0.056***
Parent Income	—	-0.012***	—	-0.013***
Parent Education	—	-0.006	—	-0.021
Parent Expectations	—	-0.130***	—	-0.113***
tanrho	0.741**	-3.138***	1.649***	2.053***
lnsigma	2.566***	2.675***	2.736***	2.758***
(*N*)	6067	1706	6067	1706
X^2	177.54	148.75	201.35	137.06
Prob X^2	0.0000	0.0000	0.000	0.000

Notes: ***probability that $B = 0 < .001$
**probability that $B = 0 < .01$
*probability that $B = 0 < .05$

Source: Author estimates Milwaukee Parental choice data base.

lower for those models, but the measures of parent income, education and expectation, more completely control for family background (and selectivity bias) for students. The MPS category is traditional schools with none of the characteristics of the specialty programs. The variables are: pat = presence of a gifted and talented program in the school (not necessarily that the student is in the program); cwp = city wide magnet program in the school; cw = city wide magnet school; p-5 = a special program focusing on basic education for schools with very poor populations.[11]

The impact of control variables is as anticipated and consistent with the value-added achievement model of prospects data presented in Table 2. Prior tests, race, gender and income all have significant and predictable effects. The grade variable indicates that as a student increases in grade, test scores relative to national norms decline. This is a common pattern in inner-city school districts. Parent education and expectations also are in the right direction and sometimes reach significant levels. The results for city wide magnet programs (cwp) and city wide magnet school (cw) are not statistically significant. Although for reading the coefficients are in the right direction, they do not meet minimal levels of statistical reliability. Math effects are close to zero.

Why the differences in the two Milwaukee studies? There are at least two plausible answers. First, the methodologies of the two studies might explain the outcomes. The first study is based on aggregate data with modest controls; the latter on student-level data and more adequate controls on student and family characteristics. The second may be in the time frame of the studies. Data for the first study were 1985, the second, 1991 to 1994. Between these studies there was a concerted effort in MPS to equalize access to magnet schools, especially middle and high schools. The main thrust was to provide minimal admission standards and then to randomly select students from within the qualified populations (taking racial balance into account). This differed considerably from the earlier period in which much more personal selection was used in admission to choice schools (Archbald, 1988; Witte & Archbald, 1985). That shift was consistent with a similar shift noted by Blank in his study from 1982 to 1989. He found "... that districts which have increased the number of magnet schools tend to avoid using selection criteria that will label the school as selective or elitist" (Blank, 1990, p. 91).[12]

Other District Magnet School Studies

Several other studies of the educational effects of magnet schools have been highlighted nationally, and more may exist that have not been circulated in research circles. Unfortunately, reliable and knowledgeable reviewers indicate that little is known with certainty. Orfield and Eaton (1996) have provided the most recent and comprehensive review. Two programs they highlight are Kansas City and Prince George's County Maryland. Both systems used extensive systems of magnet schools to reduce segregation. The Kansas City studies were done by the school district. The Orfield team review was extremely sanguine in their conclusions of the overall effects of magnets, but cited positive effects for elementary magnet schools:

> Educational achievement data are promising in some respects but the gains are modest. Although there has been an absolute increase in performance for grades K through eleven, it is unclear how students would compare to national norms. The strongest evidence for dramatic improvement seems to be confined to magnet schools at the elementary level (p. 260).

The studies of elementary schools, however, were extremely crude cohort tracking studies over time. These types of studies do not allow for changing populations or other relevant factors. Further, there were no effective controls on selectivity bias in any of the Kansas City studies.

A similar non-conclusion emerged from the Prince George's County magnet program:

> The lack of consistent and reliable empirical data and detailed program evaluations means there is little evidence with which to answer the question of whether magnet schools have improved system wide achievement compared to national norms. In general, the evidence suggest that the claims of improved achievement from magnets were based upon outdated and likely invalid measures. Unfortunately, no firm conclusion can be drawn from other available test score data (Orfield & Eaton, 1996, p. 280).

Thus, in two of the most highlighted recent programs employing magnet schools, the results, although they should not be interpreted negatively, are essentially unknown. Given the costs and turmoil of these programs, to have essentially no idea of their effects on student performance raises a major policy issue to which I will return at the end of the paper. As will be demonstrated in the following brief sections, the same general conclusion applies to inter-district choice, open enrollment, and charter schools.

INTER-DISTRICT CHOICE PROGRAMS

Although *Milliken v. Bradley* (1974) effectively foreclosed systematic integration across school district lines, especially integration of inclusive metropolitan areas, several cities voluntarily reached agreements to attempt to transfer students between city and suburban districts to enhance racial integration. Some of these transfers occurred in county-wide districts such as Louisville, and others occurred across county and district lines such as in St. Louis and Milwaukee.

The Milwaukee city-suburban transfer program, known locally as the Chapter 220 program (after the number of the state statute creating the program), is typical of these programs. The program is limited to sending racial minority students from MPS to one of 24 suburban districts and transferring white students from the suburbs to MPS. Approximately 6,000 students leave MPS each year, and about 1,000 students come to MPS from the suburban districts. Ideally, we would like to know the characteristics of those who are selected into this program and the performance of the students once they transfer.

As of 1991, we have completed a study of the characteristics of minority students going to the suburbs from MPS. This seems to be the only study of this type in the country. There is no reliable study of the achievement effects on students going in either direction, and we know of no other study elsewhere.

The study of who goes to the suburbs from MPS was based on surveys given a random sample of non-choosing MPS parents and the full set of Chapter 220 parents of transferring students. The results provided striking evidence that students transferring to the suburbs were atypical of the average MPS student in a number of important ways. In terms of race, Asian students were very much over represented in the transfer program, and African American students somewhat over represented. Socioeconomic status, parental education, parental expectations, and student prior test scores were also significantly higher for Chapter 220 students than for the average MPS student. The parents were, however, considerably less satisfied with their prior schools in MPS than the average non-choosing MPS parent (Witte & Thorn, 1996, Tables 9 and 10). Overall, the program indicates a classic example of student creaming by the suburban districts. Changes subsequent to 1991 may have altered that result to a degree (Witte & Thorn, 1996, p. 216, footnote 14).

Analysis of student achievement effects in the Milwaukee case were hampered by several factors. First, different types of achievement tests were used in a number of receiving districts, and no state testing was available at that time. Second, there was no legal way to force districts to cooperate with an objective evaluation and none offered. Finally, such an evaluation would not be easy. It would require either assessment of a range of variables to describe differences across receiving districts, or the use of a fixed effects model with quite heroic assumptions.[13] These problems can be assumed to apply to other areas and districts and may account for the paucity of systematic information on inter-district transfer programs across the country.

OPEN ENROLLMENT

Open enrollment is a state policy that in theory allows students to transfer to any district in the state. In one form or another, state funding follows the student. Since the first policy was enacted in Minnesota in 1986, however, a number of restrictions on such choices have been levied. The most common is that districts can specify quite explicitly the number, grade, and location of available slots for out-of-district transfers. This, combined, with the practical matter of transportation time and cost, has limited the use of open enrollment. Over twenty states, however, have passed open enrollment legislation.

Unfortunately, there has been extensive study of open enrollment in only one state, Minnesota. That study does not include any direct measures of the effects on student outcomes (Funkhouser & Colopy, 1994). The Minnesota study attempted to determine the characteristics of students using the open enrollment option, but the data were limited in that comparable data in Minneapolis, St. Paul, and the surrounding suburbs were not available. As for the rest of the state, students in open enrollment seemed to mirror the racial and socioeconomic composition of the remaining state districts. That, of course, tells us little since most of the racial and class diversity in the state is in the Minneapolis area.

The only measures on student outcomes were indirect and were measured by surveying the attitudes of district administrators. Although some high impact districts (either gaining or losing students) reported some changes in indirect education indicators (e.g., student teacher ratios, advanced placement courses, and high school electives), the study authors reported that "... the cause of the change in indicators is most often cited as factors other than Open Enrollment" (Funkhouser & Colopy, 1994, p. ix).

Because students will be shifting between districts and will be going in many directions, the task of systematically studying direct educational outcomes for open enrollment will be even more difficult than for interdistrict transfer programs. At least in those programs, a large initial sample could be drawn from an urban district and then tracked to a limited number of suburban districts. For open enrollment neither condition (a single large sample and a small number of receiving districts) would be present. Faced with these problems and the limited use of open enrollment, rigorous study of its effects on student achievement is probably not cost effective or a priority compared to studying the effects of the other choice options described in this paper.

CHARTER SCHOOLS

The most recent choice innovation, charter schools, falls somewhere between public and private schools depending on the form enacted in individual states. In every state there are distinctive statutes governing public and private schools. The latter are considerably less weighty than the former because most private schools are religiously based. Significant regulation of private schools would thus constitute entanglement with religion under current interpretation of the First Amendment. Charter schools are, therefore, a legal hybrid in education policy.

What is already known about charter schools is that the idea has spread very quickly from the passage of the first charter school law in Minnesota in

1991 to incorporate 25 states as of 1996. A number of studies have been done that summarize and categorize various aspects of these laws (Bierlein & Mulholland, 1994; Maugh-Pugh, 1995; Vergari & Mintrom, 1995; Wohlsetter, et al., 1995). All studies agree that the range of legislation is enormous. The common idea is that schools will be created with unique and explicit missions and/or curricular foci or methods and that the essential aspects of the schools will be provided in a charter document. Beyond that, states have created a range of options on how charter schools will be created, funded, organized, staffed, and held accountable. Some, such as those in Wisconsin, are little more than magnet schools with less reporting and regulatory requirements. In other states, such as Michigan, Arizona, and Delaware, the legislation is very permissive, and the charter schools are much closer to independent private schools than public schools.

Although a number of states have passed charter school laws and more will certainly do so in coming years, as of this writing, the total number of charter schools in existence is approximately 360 (Wells, Grutzik & Carnochan, 1996). The promise of charter schools as sources of innovation and as a mechanism to provide increased choice insures that they will be around for years, perhaps rivaling magnet schools in numbers and merging with them in some districts. They appear to be a more appealing alternative than private school vouchers for many groups. The fact that establishment school groups seem to support the movement bodes well for their survival and expansion (see e.g., Shanker, 1988).

Knowledge concerning the results of charter schools is another matter, however. What have begun are the first of undoubtedly long lists of descriptive testimonials concerning the innovative nature of these schools (Raywid, 1995; Nathan, 1994; Kolderie, 1992). These studies serve a useful purpose in that they disseminate ideas for school innovation and provide useful information for states wishing to enact or modify charter school legislation. They of course do not provide adequate evaluation of the performance outcomes of charter schools. That will hopefully be the third phase of research on the charter school movement.[14]

The problems of evaluating charter schools may be daunting, however. First, the movement encompasses very different programs across states, and by their very nature, the schools created in any single state are likely to vary enormously by level, design, and even educational purpose. For example, one common form of charter school has been as an alternative high school for students not performing well or dropping out of traditional high schools. Such schools are very difficult to judge in terms of standard measures (grades, test scores, attainment, and behavioral problems), and there is usually not a suitable comparison group of schools or students or even a theoretical counterfactual against which to test success or failure. Because, at present, charter schools are restricted in most states to just a few sites, sample sizes of common types of schools will also cause problems.

This does not mean that we should discourage efforts to evaluate charter schools in terms of performance and outcomes. Glorious stories and descriptions of imaginative and successful schools eventually run the course. If that is all that is available, the movement may go the way of other reform fads.

CONCLUSIONS

It is clear that the public school choice is a major reform movement in the last several decades. In terms of actions, rather than politics or rhetoric, it is far more important than private school choice. Even for the most researched type of choice, magnet and district choice schools, our knowledge of the effects on achievement are somewhat thin. For those categories of schools, however, what is known is generally positive. Students in choice schools seem to do better on standardized achievement tests. In some of the studies, however, unmeasured selectivity by choice schools is a problem in estimating performance effects. For the remaining choice program—inter-district transfers, open enrollment, and most recently charter schools—very little beyond anecdotal evidence is available on their effects on student performance.

Given the expansion of choice options, especially in troubled large, urban districts, more adequate research on choice and charter schools seems to be called for. More attention needs to be paid to the selection effects of various programs. Outcome effects for both choice schools, and the remaining schools need to be documented. What are the overall effects on district achievement? Who is gaining and who is declining? What are the effects on the attitudes and morale of teachers and administrators both in and not in choice and charter schools?

NOTES

1. The public school movement was dominated by Protestants and a major issue was which Bible, Protestant or Catholic, was to be used in the classroom. Church and state issues have come a long way indeed. See Tyack and Hansot (1982); Gleason (1985).

2. That program spawned a court challenge, which eventually culminated in a 5-4 Supreme Court decision, *Mueller v. Allen* (1983), supporting the program. That case remains a key precedent in the current cases for private-school voucher programs.

3. State wide alternative programs have not been emulated by other states as often as have open enrollment and charter school laws. Many alternative school choice options exist, however, in districts as magnet, charter, or private contract schools or as other options. Although they are not considered a separate public school choice program for the purposes of this paper (see Table 1), they are an important aspect of choice, especially in large school districts.

4. For a critical analysis of these types of programs, see Henig (1994, Chap. 6).

5. For a more extended discussion and analysis of the Montgomery County program, see Henig (1994, pp. 157-159).

6. These results are very consistent with the findings in the Milwaukee Parental Choice program, which is the first private school voucher program in the United States. See Witte and Thorn (1996); Witte (1997).

7. Using logit analyses, he also found that students attending private schools, Catholic and independent, were more likely to be white and from high SES homes than students in magnet school (Gamoran, 1996, p. 8, Table 33).

8. This problem was originally described for high school and beyond data in *The causal analysis of cognitive outcomes in the Coleman, Hoffer, and Kilgore Report* (Goldberger & Cain, 1982).

9. When magnet schools were included in the equations, they had large positive coefficients, but equally large standard errors due to the small number of magnet schools. In those runs, the positive coefficient in Table 2 for "comprehensive choice" was much larger (4.01) and significant at the .01 level. This probably was a function of the overlapping designation of choice and magnet schools noted earlier in the text.

10. For an interesting discussion of the creation and success of the Milwaukee magnet school program by its chief architect, see Bennett (1990). For a much more cynical description, see Archbald (1988).

11. The models also include an inverse Mills ratio (Heckman) correction for students not tested in every year. This is important because students not tested might have left the system during the 1991-1994 time frame of the study and because students tested each year are more likely to become low income students who were in Chapter 1 programs where annual testing was required. The probit for inclusion in testing indicated a disproportionate number of Black, Hispanic, low income students, and students in lower grades were tested.

12. One can speculate that the expansion of magnet schools and programs beyond the initial purpose to aid integration was based on an inherent internal contradiction in the theory of using magnet schools to enhance voluntary integration. In order to hold middle-class parents, magnet schools had to provide the type of elite educational promise associated with the original magnet schools. Once the nature of these schools became known, however, complaints about dual school systems and inequality between schools were supplemented by calls for broader choices for all parents. The answer was to expand and probably dilute the elite nature of the magnet schools and programs. The impetus for this expansion in Milwaukee came not only from minority parents and interest groups, but also from teachers and administrators in non-magnet schools (Archbald & Witte, 1985).

13. Fixed effects models include indicator variables for all units in the analysis. The idea is that they will capture any systematic, but unmeasured variance related to the dependent variable of interest.

14. In 1995, the U.S. Department of Education gave a grant to a group centered in Berkeley, California, to study the expansion and effects of charter schools. The four year study began in 1996. A report on the first phase is due to be released in 1997. Achievement results will be included in the second phase.

REFERENCES

Archbald, D. A. (1988). *Magnet schools, voluntary desegregation, and public choice theory: Limits and possibilities in a big city school system*. Ph.D. Dissertation, Department of Education Policy Studies. University of Wisconsin-Madison.

Archbald, D. A., & Witte, J. F. (1985). *Metropolitan Milwaukee specialty schools and programs* (Commission on the Quality and Equity of the Milwaukee Metropolitan Schools Rep. No. 5). Madison, WI: Department of Public Instruction.

Bennett, D. A. (1990). Choice and desegregation. In W. S. Clune & J. F. Witte (Eds.), *Choice and control in American education: The practice of choice, decentralization and school restructuring* (Vol. II, pp. 125-152). New York: Falmer Press.
Bierlein, L., & Mulholland, L. (1994). *Charter schools: A glance at the issues.* Arizona State University: Morrison Institute of Public Policy.
Blank, R. (1990). Educational effects of magnet high schools. In W. S. Clune & J. F. Witte (Eds.), *Choice and control in American education: The practice of choice, decentralization and school restructuring* (Vol II, pp. 77-110). New York: Falmer Press.
Brown v. Board of Education, 347 U.S. 483 (1954).
Driscoll, M. E. (1993). Choice, achievement , and school community. In E. Rassell & R. Rothstein (Eds.), *School choice: Examining the evidence* (pp. 147-172). Washington, DC: Economic Policy Institute.
Fliegel, S. (1993). *Miracle in East Harlem: The fight for choice in public education.* New York: The Manhattan Institute.
Funkhouser, J. E., & Colopy, K. W. (1994). *Minnesota's open enrollment option: Impact on school districts.* Washington, DC: Policy Studies Associates.
Gamaron, A. (1996). Student achievement in public magnet, public comprehensive, and private city high schools. *Educational Evaluation and Policy Analysis, 18* (1), 1-18.
Goldberger, A. S., & Cain, G. G. (1982). The causal analysis of cognitive outcomes in the Coleman, Hoffer, and Kilgore Report. *Sociology of Education, 55*, 103-122.
Henig, J. R. (1994). *Rethinking school choice: Limits of the market metaphor.* Princeton, NJ: Princeton University Press.
Kolderie, T. (1992). Chartering Diversity. *Equity and Choice, 9*, 28- 31.
Maughs-Pugh, T. (1995, July). Charter Schools, 1995: A survey and analysis of the laws and practices of the states. In *Online Education Policy Analysis Archives.* Available: Doc. No. 3.
Millikan v. Bradley, 418 U.S. 717 (1974).
Moore, D. R., & Davenport, S. (1988). *The new improved sorting machine.* Madison, WI: National Center on Effective Secondary Schools.
Mueller v. Allen, 463 U.S. 388 (1983).
Nathan, J. (1994). *Charter public schools: A brief history and preliminary lessons.* Minneapolis: Center for School Change, Hubert H. Humphrey Institute of Public Policy.
Orfield, G., & Eaton, S. E. (1996). *Dismantling desegregation: The quiet reversal of Brown v. Board of Education.* New York: The New Press.
Plank, S., Schiller, K. S., Schneider, B., & Coleman, J. S. (1993). Effects of choice in education. In E. Rassell & R. Rothstein (Eds.), *School Choice: Examining the Evidence* (pp. 111-134). Washington, DC: Economic Policy Institute:
Raywid, M. A. (1995). The struggles and joys of trailblazing: A tale of two charter schools. *Phi Delta Kappan, 76* (7), 555-560.
Rossell, C. H. (1990). *The carrot or the stick for school desegregation policy: Magnetschools or forced busing.* Philadelphia: Temple University Press.
Shanker, A. (1988, July 10). A Charter for Change. *New York Times*, (Reprinted by the American Federation of Teachers).
Tyack, D., & Hansot, E. (1982). *Managers of virtue.* New York: Basic Books.
Vergari, S., & Mintrom, M. (1995). *Charter school laws across the United States.* Lansing, MI. Institute for Public Policy and Social Research.
Wells, A. S., Grutzik, C., & Carnochan, S. (1996). *The multiple meanings of U.S. charter school reform: Exploring the politics of deregulation.* Paper presented at the British Education Research Association Meetings, Lancaster, U.K., September 12-15.
Witte, J. F. (1992). Private versus public school achievement: Are there findings that should affect the educational choice debate? *Economics of Education Review, 11* (4), 371-394.

Witte, J. F. (1997). *Achievement effects of the Milwaukee voucher program. Paper presented at the American Economic Association Annual Meeting, New Orleans, January 4-7.*
Witte, J. F., & Archbald, D. [1985]. Metropolitan Milwaukee specialty schools and programs. Milwaukee, WI: Report 5: Study Commission on the quality and Equity in the Metropolitan Milwaukee Public Schools.
Witte, J. F., & Thorn, C. A. (1996). Who chooses? Voucher and interdistrict choice programs in Milwaukee. *American Journal of Education, 104* (3), 186-217.
Witte, J. F., & Walsh, D. J. (1990). A systematic test of the effective schools model. *Educational Evaluation and Policy Analysis, 12* (2), 188-212.
Wohlsetter, P. (1995). Charter schools in the United States: The question of autonomy. *Educational Policy, 9*, 331-58.

ACCOUNTABILITY UNDER CHARTERS AND OTHER SCHOOL-CENTERED REFORMS

Paul T. Hill

Various reform proposals would transfer control of public funds from centralized bureaucracies to individual schools, encourage creation of systems of distinctive schools, and give families choices among public schools. Though these proposals differ in many ways, they are all school-centered. They all try to strengthen local schools as organizations and hold schools as whole organizations accountable for student learning.

School-centered proposals take several forms, including school system decentralization, charter school laws, and public school contracting. Decentralization decreases regulation of schools, increases school control of funds, and emphasizes accountability for performance, not compliance. Charter schools laws permit schools to become self managed, receive public funds on a per pupil basis, control staffing and spending, enroll students via family choice, and be held accountable only for student performance. Under public school contracting, schools would be operated by independent

Advances in Educational Administration, Volume 5, pages 191-207.

ISBN: 0-7623-0081-7

organizations under explicit and legally enforceable agreements with school boards, funded on a per pupil basis depending on family choice, staffed only by teachers chosen by the school, and held accountable only for meeting the terms of their contracts.[1]

These proposals imply a radical yet plausible answer to the question, "How can schools be public yet effective?" They assume public education can be improved by making schools real organizations, in control of their funds, staffs, and programs, and that competition, innovation, and differentiation on the basis of approach to instruction can strengthen, not weaken schools. Such proposals arouse profound opposition from groups that equate "public" with "government controlled," and think "equity" requires literally identical educational experiences for all students. Because opponents have worked to keep such proposals off the official agenda, these ideas have never been tried in their pure forms. Even school system decentralization, which has at least the nominal support of school boards and teachers' unions in cities like Chicago, Denver, Seattle, Los Angeles, and Cincinnati, has been implemented only in part, due to insiders' reluctance to give schools full control of dollars and teacher hiring and firing.[2]

Aside from the political issues surrounding the enactment and implementation of school-centered reforms, there are major unresolved questions about how they would operate. Chief among those issues is accountability: How, under a system of support for public education that would place control of funds, resources, and methods at the school level, would individual schools be held accountable for performance? This paper lays the groundwork for an answer to that question. It discusses the accountability issues that school-centered reform brings to the fore, and it shows how these issues might be resolved by a mixture of parental choice, school specific agreements about school mission and performance expectations, value-added achievement testing,[3] and more authentic measures of student outcomes. It also suggests how the accountability arrangements could be applied under a variety of reforms, including school system decentralization, public school contracting, and charter schools.

ACCOUNTABILITY ISSUES RAISED BY SCHOOL-CENTERED REFORM

Decentralization and other school-centered reforms revolutionize accountability in K-12 public education because they focus on the school as a whole organization, rather than on individual components of the school or on the broader local system. Each school will have its own approach to instruction, its own decision making processes, and control of its own resources. The school board will oversee the school as a whole, and it will neither micromanage the staff nor control the separate instructional programs that make up a school. Thus, the school board will not be concerned about whether a given teacher or instructional program is productive. It will only be concerned about whether the school as a whole serves its students well.

School-centered reform implies three profound changes in accountability:

1. It eliminates the bureaucratic practice of holding parts of schools separately accountable—different categorical program coordinators responsible for compliance with different sets of regulations, the principal responsible for compliance with provisions of the union contract and district reporting requirements, and individual teachers responsible for fulfilling curriculum coverage requirements—and establishes joint schoolwide accountability for overall performance.
2. It changes what schools are accountable for, eliminating the bureaucratic practices of holding school people accountable for compliance and paying little attention to the school's overall reputation or productivity.
3. It changes whom schools are accountable to, replacing a system based on reporting upward within an administrative hierarchy with a more complex system in which community wide goals for student achievement, professional standards, and family preferences all matter.

Whole School Accountability

School-centered reform changes the terms of employment for individual teachers and administrators. Their terms and conditions of employment will be determined by the instructional philosophies and needs of the schools in which they work.[4] Though some will have rights of transfer to other public schools, their jobs will be defined more by the individual school's mission and student needs than by district wide work rules. In the case of charter and contract schools, teachers and administrators might work for individual schools and have no formal employment relationship with the school district. Under these circumstances, teachers' and administrators' jobs will depend heavily on the performance and reputation of the schools in which they work.

In contrast, accountability in conventional school systems focuses mainly on individuals and on programs, and far less on schools. Districts run civil service systems for administrators and teachers, and make most decisions about pay and promotion on the basis of personal characteristics—chiefly training and experience. Mediocre performers enjoy the same pay and job security as do above average performers, and the only real threats to jobs and raises are personal scandal or financial impropriety. Outstanding or egregious individuals can be reassigned on the basis of their reputations, but the vast majority are paid for their attributes, not performance. On the program side, districts normally assess effectiveness on a district wide basis rather than school by school. Even in districts that publish school report cards, the traditional civil service system channels most rewards and penalties to individuals and programs, not to schools as wholes. Only a few districts (e.g., Charlotte, North Carolina) monitor student achievement gains on a school by school basis and administer important rewards and penalties (including hiring and firing of staff members and possible school closing) on the basis of overall school performance.

Accountability Emphasizing Performance

All forms of school-centered reform involve some agreement between the local school board or superintendent and the individual school about goals, basic modes of operation, problems to be remedied, sources of assistance, and performance requirements. These agreements are more explicit and legally enforceable under charter school and educational contracting laws than under decentralization, but some form of agreement is necessary.[5] The agreement is the quintessential accountability mechanism.[6] It establishes the district's right to expect the school to admit students by known rules and criteria, to serve all students enrolled, and to attain specified student outcomes. The agreement also establishes the school's right to operate without interference, receive specific forms of assistance, be paid for pupils enrolled, be free to enroll students from a given catchment area, recruit teachers, and have its agreement continued or renewed if it meets all specified criteria.

Schools can be accountable for five kinds of outcomes:

1. Compliance: whether the school (or a part of it) abides by applicable laws, regulations, policies, contracts, and grant conditions, including, when applicable, teacher union contracts and rules for use of state, federal, and foundation grant funds.
2. Controversy: whether the school (or a part of it) avoids improper or notorious conduct, related to use of public funds, treatment of children, personal deportment.
3. Student performance: whether the students in the school exhibit gains on tests or other measuring instruments intended to gauge their level of mastery of facts and broader subject matters, or their ability to apply knowledge.[7]
4. Ultimate student results: whether students, having left the school, are able to perform as required at the next level or education or in some real world context (e.g., jobs, the military).
5. Family satisfaction: whether families are satisfied with their children's school experience (e.g., as measured on a survey) or confident enough that the school is meeting their children's needs to choose their children's current schools over other local public schools.

Table 1. Weights Attached to Different Outcomes by the Current System and by School-centered Reform

	Compliance	*Controversy*	*Student Performance*	*Ultimate Student Results*	*Family Satisfaction*
Centralized System	Highest	High	Low	Low	Lowest
School centered Reform	Lowest	Low	Highest	High	High

As Table 1 illustrates, school centered reform weights these outcomes very differently than does the traditional centralized system.

Though the people who supervise and operate the traditional centralized system are personally deeply concerned about student performance, they can seldom act on it at the school level. As Table 2 shows there is seldom much information about outcomes other than compliance and controversy, for the very good reason that boards have a very hard time using such information. A school that is in compliance and is not controversial because of a moral or safety issue cannot be changed much. Though some districts do survey parents and publish school by school achievement results, the existing system is not structured to respond strongly and regularly to these outcomes.

Since the publication of Friedman's (1962) first education voucher proposal, analysts have noted that government's effort to improve public education by creating successive layers of regulation has made schools into compliance organizations. School-centered reform is a way of restoring student results to the center of the accountability process.

Accountability to Families and Professional Interests as Well as to Government

Decentralization and other forms of school-centered reform can make schools accountable both to the broader community, which funds schools and

Table 2. Current Availability of Information on School level Outcomes

	Compliance	*Controversy*	*School specific Student Performance*	*Ultimate Student Results*	*Family Satisfaction*
Current availability of relevant information	Readily available	Readily available	Seldom available	Not available	Seldom available

compels student attendance in order to guarantee its future social, economic, and political viability, and to families, which have their own distinct aspirations and responsibilities for children's education; and to professional standards.

Family Interests

School-centered reform seeks to restore a balance between the private and public interests in education. In recent years, urban parents have found public schools unresponsive and for good reason. Schools, especially in big cities, are so constrained by rules and regulations made elsewhere—governing, for example, how students and teachers are selected, what teachers can help what students, how long a child must be kept out of regular classes in order to receive bilingual education, what a principal must go through in order to remove a disruptive student, and what work teachers may and may not do—that many are unable to respond to the needs of students and parents. As Ted Sizer has said, parents experiencing the blank bureaucratic face of such schools have reason to question whether they are really public at all: They appear profoundly private, controlled by government on behalf of teachers' unions and interest groups that have managed to capture the agendas of school boards and state legislatures.

School-centered reform tries to restore the partnership between schools and families that has been destroyed during a period of regulatory excess. It recognizes the private interest in schooling by allowing families to choose among schools that take different approaches to education.

Public Interest

School-centered reform recognizes the public interest in schooling, ensuring that students learn basic skills, prepare for responsible lives as earners and citizens, and understand basic democratic values. To protect these public interests, it requires that every school operate under an agreement with a state agency, the school board, and be accountable for performance. Outcomes for which a particular school is accountable can be linked to national or state standards or can be tailor made for a particular kind of school. Career oriented schools, for example, aspire to outcomes that regular academic schools do not. Schools that do not provide instruction as promised, or whose students' outcomes are low and not improving as anticipated, can be closed or reorganized.

Teacher and Administrator Interests

Under school-centered reform, schools must attract administrators and teachers. Though schools will also have choices about whom they employ, their capacity to hire and maintain high quality staffs will depend on pay, working conditions, and institutional reputation. Under such conditions, schools will need to maintain their reputations as secure and satisfying places to work. Potential staff members might assess schools according to criteria similar to those used by parents or the school board. Schools based on weak ideas or burdened by poor management, however, might be sanctioned by the departure of essential staff members even sooner than they are abandoned by parents and the school board. Similarly, schools that gain good reputations among teachers will probably gain parental and board support due to positive mentions by teachers.

Professional Standards

Many schools operating under school centered reforms will join voluntary networks of like-minded schools for the purpose of sharing knowledge and making efficient joint investments in teacher training and assessment. Some of these might operate as inspectorates, as do some private school networks, alerting their managers and boards to declines in instructional quality (Celio, 1995). As Bryk (1997) has suggested, school districts and other school authorizing agencies might also fulfill their oversight responsibilities by creating inspectorates intended to identify school performance problems and identify charter schools that need help or other intervention.

Enterprise Interests

School managers will be accountable to their site councils or boards of directors for fiscal soundness, instructional quality, and fidelity with the school mission. These interests should promote overall school quality and should reinforce, not conflict with, parent, board, and professional interests. A school that is one among several overseen by a larger public service organization (e.g., Seattle's Powerful Schools) is also accountable for program quality and fidelity to the central management of those organizations.[8] These multi-school organizations will also be concerned about the quality and consistency of school instructional programs. Like existing private school networks, multi-school organizations might occasionally weaken a particular school by drawing away a talented staff member needed to rescue another school. The broader organization's concern for its reputation and relationships with school boards and parents, however, motivates some vigilance about quality control and problem solving.

Under school-centered reform, the primary mechanisms of accountability are two: schools' agreements with school boards and parental choice. Secondary mechanisms include professional labor markets and enterprise quality control. Table 3 displays the advantages and problems of the different accountability mechanisms. None is sufficient alone, but together they provide a more balanced accountability system than is now available in public education.

Table 3. Mechanisms of Accountability under School-centered Reform

	Mechanism	*Direction*	*Advantages*	*Problems*
Agreement with school board	Students must meet learning gain requirements negotiated between the school operator and the school board. School must also comply with the terms of its contract (e.g., on student admission and instruction focus).	Upward: All accountability is to governmental agency.	Emphasizes the importance of student achievement gains indexed on broader community, state, or national standards. Gives the board strong action options in case of noncompliance.	Tests cover only a sample of skills; pressure to teach to the tests can narrow the curriculum. Raw test scores do not reflect school's value added. Compliance does not necessarily reflect student benefits.
Family choice	School must attract families. Schools that fail to attract families lose funding and close.	Downward: All accountability is to families who act by exercising choice.	Ignores accountability to government agencies, broader standards.	Vulnerable to school distortion of facts, misplaced parental confidence.
Teacher and administrator support	Schools must attract and keep competent teaching staff.	Laterally, to teachers who have choices about where they work.	Reflects school's ability to develop support among teachers.	Creates severe problems for schools that cannot provide good teacher working conditions.

Professional standards	Schools must win accreditation and maintain respect of professionals.	Laterally, to the professional standards and teacher opinion.	Emphasizes quality of school climate and teachers' sense of productivity.	Vulnerable to professional interests in salary, benefits, pay for time not worked.
Enterprise management	School managers must answer to boards of directors and sometimes to top managers of multiple school organizations.	Upward within the school enterprise to boards of directors and network managers.	Emphasizes fidelity to school mission; intermediary organization accountability can bring expert resources from other schools.	Vulnerable to splits on boards of directors, conflicts between individual and multi-school organization interests.

Together, they place the school where it should be, answering to the broader community, as represented by the school board, families, professionals, and enterprise managers.

If a school fails to meet agreed upon student achievement targets, or if it violates its promises (e.g., by "skimming" student applicants instead of picking at random, or offering a watered down curriculum instead of the Montessori program promised), the agreement can be terminated. Similarly, a school that cannot attract enough students to meet its costs or keep enough good teachers must close. Parents, school boards, and teachers can exercise their options, however, only if they have other alternatives. Boards, for example, must know of other competent providers who can take responsibility for a failing school, parents must know of other schools that have space for their children, and professionals must work in a labor market where many schools have discretion about whom they will hire.[9]

On the positive side, the staff and community supporters of a highly successful school can be rewarded by the security of their school's status and, on occasion, by the opportunity to expand their school (e.g., by adding grades) to lead multi-school networks or to assist or manage other schools. Not all school communities will want to expand or take responsibility for additional schools, but some might. Schools that do not expand will still be rewarded by their ability to attract students and high quality teachers and administrators.

Penalties and rewards are fundamental to school-centered reform. Aggressive use of their powers to enter performance agreements with individual schools can make school boards much more effective. Using the school centered reform process to eliminate weak school providers and replace them with more capable ones will raise overall local performance standards and stimulate a strong supply response by individuals and organizations that have definite ideas about schooling and can demonstrate the capacity to attract and train capable teachers, engage students and parents, and run coherent and productive schools.

The remainder of this paper explores school-centered reform's accountability relationships in greater detail.

TOWARD A SCHOOL-FOCUSED PUBLIC ACCOUNTABILITY SYSTEM

When school-centered reform is in full operation a school can be held accountable to the local public school board in five different ways. First, it will be directly accountable to the school board for compliance with the terms of its agreement—providing the type of instruction promised, offering classes the promised number of days, employing a staff that meets the description provided in the agreement, involving parents as promised, admitting and expelling students only on the grounds specified, and so on. Second, a school will be accountable to the school board for immediate student outcomes—Do student gain scores match the expectations agreed to in advance, and do all groups of students in the school benefit as expected? Third, a mature school, one that has sent one or more cohorts of students into higher levels of education or into the labor force, can be accountable for the performance of its graduates. If agreements between the district and the schools anticipate this possibility, elementary schools can be held accountable for the proportions of their graduates who are able to pass key secondary school gatekeeper courses (e.g., algebra), just as high schools can be rated on the proportions of their students able to gain admission to four year colleges, maintain passing grades, and graduate within a stated period. Fourth, a school can be held accountable for maintaining high ratings for professional quality, as reached by members of an inspectorate or through an accreditation process.

The fourth basis of accountability illustrates how the contract system will operate in practice. A school board deciding whether to continue or renew a charter or other agreement with a school that has met most but not all expectations (e.g., student test score gains are lower than anticipated but still above the district wide average) should want to seek expert advice about whether the school curriculum, instruction, climate, and teacher training meet high professional standards.[10] A board considering a school with a good but imperfect record will have to make practical judgments about whether the contractor has diligently worked to maintain quality, performed well in similar circumstances, and has the capacity to improve. An independent inspectorate, possibly one created to assess the quality of schools of a certain type, can provide such judgment credibly.

None of the four bases of accountability is perfect, and each poses challenges of two kinds: technical challenges, associated with making valid measures of the relevant outcome, and integrity challenges, associated with ensuring that neither individual schools nor their competitors or detractors can bias the ways outcomes are assessed. Table 4 summarizes these challenges and suggests likely solutions.

One technical challenge of great importance is the need to reflect school specific value added on student achievement. Agreements must set ambitious goals for student achievement gains, but they must not create incentives for school staffs to avoid low income students or others whose absolute achievement levels start out below average. A school board that entered agreements requiring all schools to meet particular absolute achievement levels could place insupportable burdens on schools that, because of their location in low income areas or their chosen mission, served educationally disadvantaged children. Such schools might be targeted for penalties or closure even if their students gained at a high rate. Families served by such schools could find that promising schools were constantly being snatched away from them and replaced by unfamiliar new start-ups.

Table 4. Challenges of School-Specific Accountability and Possible Solutions

	Compliance with school board agreement	*Agreement provisions re: student achievement*	*Ultimate student results*	*Expert judgment, accreditation*
Technical Challenges	Establishment of objectively measurable indicators of compliance (e.g., whether the school offered the curriculum promised or admitted students fairly).	Coverage of all important learning areas; psychometric validity and reliability; freedom from cultural bias; need to reflect school-specific value added.	Need to follow up samples of graduates into other institutions. High cost of follow-up. Difficulty finding students from high-mobility schools.	Gaining access to organizations able to judge or accredit schools of a particular type.
Integrity Challenges	Possible dilution of instructional methods of standards by school staff. "Sticklerish" enforcement by a hostile board.	Possible school meddling with test result or manipulation of who is tested.	Possible distortion by schools. Need to ensure high response rate and reflect a range of student futures beyond higher education.	Possible "sweetheart" relationships between school operators and accreditors.
Likely Solutions	School's right of appeal to state board or other neutral party on interpretation of agreement.	"Secure" testing by independent parties. Value added assessment. Gain score comparisons among schools serving similar populations.	Data collection, analysis, and publication all done by an independent contractor paid by local school board.	Possibly a publicly paid state or local inspectorate or private organizations capable of accrediting a variety of school types.

Under school-centered reform, school boards must know whether disadvantaged students in a school are learning at a rate that closes the gap between their own performance and broader community standards. They must also know whether disadvantaged students' progress in a given school is high or low, relative to similar students' progress in other schools. Without that information and some ability to formulate school specific agreements that take account of value added for disadvantaged students, a school board could find itself in the tragic position of closing or reregulating a school that was doing a better job for such students than any other in the district.

State or local student achievement standards and aligned student testing systems can be important elements of school-centered reform. But decentralization and charter or contract arrangements force school boards to take a subtler position than is represented by the popular slogan, "All students must meet standards." All students should be assessed relative to standards, and schools must work to help the disadvantaged catch up. But an accountability system that automatically downgrades all schools whose entering students achieve below the age-appropriate standards will force prudent teachers and administrators to avoid challenging situations.

School boards need to reach agreements with individual schools that set high but realistic expectations. Comparison and competition among schools can force school leaders to set ambitious goals for themselves. But neither the district nor the school is well served by a situation in which school leaders must make unrealistic promises in order to gain freedom to operate. Standards for value added must reflect the best demonstrated experience of real schools, not aspirations unalloyed by experience. Responsible school boards will, on occasion, be criticized for setting lower short-term goals for disadvantaged students than for others. Rigorous value added measurement of student progress can ease, if not eliminate, boards' burdens in such situations.[11]

Another idea that surfaces in Table 4 is the importance of ultimate student results. School-centered reform gives school boards the opportunity to consider truly "authentic" assessments (i.e., of graduates' demonstrated performance in the real world) in settings that cannot be controlled or manipulated by teachers or anyone else who has a stake in a school's looking good. Unlike achievement tests (even so-called authentic or performance based examinations), these measures do not simulate real world outcomes; they are such outcomes. For high schools, authentic outcomes might include graduates' rates of admission to four year colleges, college grade points, persistence in college, graduation rates, average years to graduation, and so on. Some university admissions offices, including the University of Washington, rate feeder high schools in terms of the discrepancy between students' high school and college grade point averages. School boards could obtain such data and other authentic performance measures from cooperating colleges and universities. They might also conduct follow-up surveys to ascertain graduates' employment status, occupations, and rating of their own high school preparation.

For elementary and junior high schools, local boards could assess their graduates' ability to pass gatekeeper courses like algebra, rates of high school credit accumulation, ability to pass state proficiency examinations, and so forth. Such data will be imperfect, and differences in schools' results can reflect transitory economic conditions as well as students' incomes and family backgrounds. For these reasons, they are probably valid bases of comparison only among demographically matched schools. Despite their deficiencies, however, such authentic assessments are the best way to make it clear that the purpose of schooling is to prepare students for complex real lives, not for taking tests.

Education is not an exact science, and neither is school accountability. Any form of public education, whether one based on school-centered reform or on direct bureaucratic operation of schools, requires the use of human judgment. Decisions to reward or punish a school, to open a new school or close an old one must all be taken in light of the challenges faced and the options available. As in other complex fields, the quality of raw materials and the exact conditions of performance can never be fully anticipated in advance. School-centered reform permits school boards to consider a wide range of indicators of school performance and to weigh them in ways that are sensitive to the challenges encountered.

As in other fields, such complexity must affect the ways agreements between school boards and individual schools are written and enforced, but they do not make school centered reform itself inappropriate. School-centered reform allows a government agent, a public school board, to sponsor many different schools, each of which takes a distinctive approach to solving a challenging problem. In other fields where government is reasonably successful at overcoming technical uncertainty (e.g., the space program or new weapons development), such complex judgments must be made. In those fields, the most unstructured and uncertain tasks are done under contracts (i.e., performance agreements with independent organizations) because such agreements allow multiple approaches to a problem and because government can more readily abandon a failed line of work than it can fire civil servants or close down a federal installation. The space and defense analogy does not prove that school-centered reform is the best approach to public education, but it does prove that a system of contractual accountability can be compatible with complexity and technical uncertainly.

Table 5 addresses the question of consequences: What can happen to a charter school if it does especially well or poorly? It shows how a decentralized system creates new options for school boards, parents, and teachers to act on their judgments of school quality. As it illustrates, decentralization can create powerful new options for all parties in response to both school success

Table 5. How Different Actors Can Respond to Information about School Performance

	Possible Responses to Poor Performance	*Possible Responses to Good Performance*
School board (with superintendent and central office)	Offers of assistance; probation; forced acceptance of help; partial re staffing; allowing students to enroll elsewhere; creation of competing schools (e.g., charters); closing and reconstruction.	Awards and recognition; opportunities to assist others; opportunities for key staff members to start similar schools; extra income opportunities for staff members who work with other schools.
Teachers	Demands for new leadership; transfer out of the school.	Confidence in school leadership and enhanced cooperation; efforts by high-quality teachers to transfer into the school.
Parents	Demands for new leadership, teaching staff, and curriculum; transfer out of the school.	Support for teachers and school leadership; cooperation with school's academic demands on students; efforts by parents to enroll students in the school.

and failure. In many ways, the most important new options are in the positive column: Under decentralization school boards, teachers, and parents can reward good performance just as strongly as they can punish failure.

Table 5 also makes it clear that decentralization does not mean abandonment of public oversight of schools. Schools are free as long as they perform, but public officials are responsible to intervene in schools that fail children. As the table also illustrates, however, decentralization defines the goals and methods of public intervention. School boards, central offices, and superintendents must intervene to put schools back on their feet or to replace failing schools with new ones that can operate effectively under conditions of autonomy. They must not consign any school to the status of a permanent board "project," or reregulate all schools in the course of intervening in the most troubled ones.

CONCLUSION

The ideas about school accountability in this paper will be unfamiliar to many readers. Some might think the accountability challenges are too great to face, but they must be faced. The accountability challenges considered here, however, will arise in the course of any reform that tries to focus initiative and accountability at the school level. Any such reform requires district leaders to ask whether a school is performing as expected. Any community attempting such a reform without considering how to replace a system of compliance based accountability with one based on school specific performance will soon discover that it faces questions and choices on which it has no relevant data. The only alternative to working toward a school-specific accountability system is to retreat to a conventional system under which schools are neither free nor accountable.

This paper is an inevitably incomplete first effort to sketch out the requirements for school specific accountability. Analysts and reform sponsors interested in decentralization, charter like arrangements for schools, or school contracting need to invest time and effort in developing data systems and analysis capabilities that would support the forms of accountability sketched above.

NOTES

1. For an extended discussion of the differences among these proposals see Hill, Pierce and Guthrie (1997).

2. For elaboration on this point, see the report of a multi-university team led by the present author and Anthony Bryk (1996).

3. Many of the ideas presented here are inspired by an unpublished paper by the late James S. Coleman.

4. The new "trust agreement" between the Seattle Education Association and school board is one approach to creating this situation.

5. See Hill, Pierce and Guthrie (1997) about the advantages of formal contracts between legally independent parties. See also Millot (1996).

6. The decentralization plan in Charlotte was based on such agreements from the beginning. Chicago's decentralization plan did not provide for such agreements, but it is evolving to include them, as many schools prove unable to change without external pressure and assistance. Seattle's reform plan now includes annual agreements among the school staff and community on one hand and the superintendent and teachers' union head on the other.

7. Note that the foregoing list does not use the terms "performance-based" or "authentic assessment." It is not yet clear whether such tests, which require extensive observation and rating of students while performing tasks that imitate real world activities, will ever become sufficiently economical or reliable to be used for school accountability. The amount of student and observer time required for performance based tests is so great that it is often impossible to observe a very wide range of any one student's behavior. Widespread use of performance based tests could have the paradoxical effect of narrowing, not broadening, the range of outcomes for which schools are held accountable. Further, though performance based tests now available apparently do measure skills other than those tested by standard norm referenced examinations, it is not clear whether performance based tests have as much predictive validity (i.e., the capacity to predict a student's performance at the next higher level of education or in a job) as do norm referenced instruments. See, for example, Stecher and Koretz (1996), and Mitchell (1996). At this time, it is not clear that there will ever be a form of authentic assessment, other than observation of students' actual performance after leaving a school. This point is developed in greater detail below.

8. See Hill, Pierce and Guthrie (1997) for a discussion of the role that such organizations, called Management and Assistance Providers (MAPs), can play in quality control and accountability.

9. The contracting system can be effective in eliminating or changing weak schools only if there are alternative suppliers. The existence of intermediary organizations would increase local school boards' options. The board could assign a failing school to an organization with a good track record. Some problem schools might even be shared among intermediary organizations as

assigned risks. See Hill, Pierce and Guthrie (1997) for a discussion of the state role in ensuring that there is a market.

10. Some of the ideas presented here are based on work done for Chicago's school reform, which decentralizes some decisions to the school level but does not create real contracts between individual schools and the school board. For a discussion of ideas for accountability arising from the Chicago effort see Bryk (1997).

11. See Hill, Pierce and Guthrie (1997) for a more extensive discussion of the need for value-added measures in school contracting.

REFERENCES

Bryk, A. (1997). *Democratic localism: A lever for institutional renewal.* Los Angeles: Westview Press.

Celio, M. B. (1995). *Building and maintaining systems of schools: Lessons from religious order school networks.* Working papers of the University of Washington Graduate School of Public Affairs, Seattle.

Friedman, M. (1962). *Capitalism and freedom.* Chicago: University of Chicago Press.

Hill, P. T., & Bryk, A. (1996). *Toward a system of schools decentralization in six urban school districts.* Seattle: The University of Washington Graduate School of Public Affairs.

Hill, P. T., Pierce, L., & Guthrie, J. (1997). *Reinventing public education: How contract schools can improve American education.* Chicago: University of Chicago Press.

Stecher, B. M., & Koretz, D. (1996). *Issues in building an indicator system for mathematics and science education.* Santa Monica, CA: Rand Corp.

Millot, M. D. (1996). *Autonomy, accountability, and the values of public education.* Seattle: University of Washington Graduate School of Public Affairs.

Mitchell, K. (1996). *Norming and Conforming.* Santa Monica, CA: Rand Corporation.

RECONNECTING THE LITTLE RED SCHOOL HOUSE TO THE COMMUNITY

Sherry R. Eagle

INTRODUCTION

Bringing about change requires seeing clearly what others do not yet perceive. To move forward we must see paths and obstacles that may or may not allow a new patterns to occur. The key to change is vision, seeing a familiar image and altering it by shifting its pattern, finding new connections, or creating new expectations. This is like looking through a kaleidoscope. A kaleidoscope permits you to see multiple reflections of a given image. A kaleidoscope provides variations on the original pattern.

Reformers in any field must be able to see existing patterns in different ways, and they must be able to discuss how and why a new image is an improvement. Even the slightest turn of a kaleidoscope can change an image; one need not alter an existing pattern very much to create a significant difference.

Advances in Educational Administration, Volume 5, pages 209-217.

ISBN: 0-7623-0081-7

PERSONAL REFLECTION

Based on this kaleidoscope approach, I was anxious, as an educational practitioner, to read and react to what university professors are saying about school reform and school accountability. I readied myself by pulling out papers that I had written during my doctoral work on school choice and papers on the Tiebout theory that families of means have always had school choice. I even did an ERIC search on school choice, as it related to student performance. I was ready for the opportunity to voice my opinion to fellow practitioners on whatever the professors did not elect to elaborate in their presentations.

What happened then was surprising. I read the papers by Paul Hill and John Witte, underlined important points, wrote notes in the margin, tried to tie their ideas to practice in our school system and practices that I have heard about in other systems. The content of the papers, however, kept playing with my mind. It wasn't just the specific definitions or the premises; it was also the overall "Ah ha!" that followed my attempt to understand what these papers meant.

The shift came for me in the form of questions: Why was I ready to support the existing system? Why was I quickly ready to find fault with any reform initiative that could lead to change in the current system's structure? The reason is quite simple. The existing structure is safe. It has an order about it. Many communities accept it as positive for students. The state, by recognizing many schools' exemplary programs, continues to shower praise on programs within the current system. At the same time, other educators, business leaders, and social service and community members claim that the system is failing the youth of today. There are also those who simply say that no matter what the system does, true improvement in student achievement can only occur if a family is engaged in their child's learning and if there are significant adults in that child's life.

It is easy to see why educators continue to defend the current system; it tends to work for the majority of community families who have a strong interest in their children's education. The papers by Hill and Witte have made it apparent that an educational system cannot continue to condone unacceptable performance. If John Dewey was correct in assuming that what the brightest parents in a community want for their child is what every member of the community should want for their children, then change must happen because there are families in our communities who are not believing in the system and there are students in our communities who are not learning within our current structure. The school that is not connected to the community will not long survive.

Since reading these two papers, I have come to see that something new, such as a charter school or contract school, that can ride the tide of reform by shifting current practices and procedures while remaining part of a safe system with community driven common practices and beliefs, may provide a viable means for school systems to begin looking at alternative organizational structures.

Parents, given the opportunity of selection, may be provided with a school of choice that has as its framework the needs of the families who elect to attend. The place may look and be different from the current neighborhood school. To improve a system that is not serving current needs is like continuing to improve upon a milk delivery wagon that is no longer needed to deliver milk. Is it the children who walk through the school house door, or is it the community's support for programs and teachers inside the school house that make a high performing school?

SCHOOL-CENTERED REFORM AND ACCOUNTABILITY

The premise of Hill's paper is that the school house is the unit of analysis, and all measures of accountability must be aligned to community standards for learning. His paper discusses three options: school system decentralization, charter school laws, and contracting out for public schools. Hill assumes that if schools are given full autonomy and control over programs, staffing, and funds, improvement of student achievement will occur. He notes that if school systems begin to run like other organizations, they may decide not to act as managers in some areas and may contract out responsibilities. Hill also discusses educational contracting as an organizational means that aligns with accountability and advocates parental choice and authentic measures. The following outlines my understanding of Hill's view of contract schools.

Contract School Components

The concept of contracting has as its foundation a legal contract that defines a school's mission, its funding and accountability measures, the students it will serve, and various compliance issues (e.g., type of instruction, student outcomes, graduate performance, and contract operation) for the local school house. Schools could be operated by public and private organizations. Governance would come from the local school board, but operations could be privatized. The school could set its vision and expectations on the basis of the school community's beliefs and their ability to pay.

The concept of contracting aligns with the belief that once schools are totally managed on site, accountability rests with the teachers and parents. Contract schools could operate under federal and state guidelines; the governing board, however, would not necessarily have to build or equip them. The cost of maintenance, operations, and infrastructure would depend on the school's overall enrollment needs. Therefore, their instructional emphasis, admissions practices, student outcomes, sources of funding, and enrollment served, would all be delineated in the contract.

Contracting service provided for the public school could be any private or public non-profit or profit making organization. Contractors could have the right to use public school buildings at no cost as part of the school district's responsibility. Public funds used to support contractors could be allocated on a per pupil basis with a foundation level that would be raised due to increased enrollment or needs of at risk youth. The increased allocations for the at risk population could come from categorical grants that now offset district costs of operation, as opposed to being added for increased services. This funding mechanism would assure the same fiscal resources for all schools. At present, because of salary schedules and staff makeup, this does not happen. In the current system, per pupil expenditures vary from building to building, depending on seniority of staff.

The School's Role in Contracting

Because the school would be the central contracting agent, each would draft its own contract to include mission, educational plan, and expectations for teaching and learning. The teaching staff would be hired and evaluated by the independent contractor. Compensation would be commensurate with performance, as defined by the contracting group. This does not abandon the union concept since all teachers could be hired from union ranks. Given a specific operating budget, however, no school could afford only highly paid teachers, which may be the case now.

The Students' Role in Contracting

Student selection into contracting schools would involve random selection from lists of applicants. This would insure the public aspect of schools. Students in this situation are selected, not admitted. Once selected, however, students would be subject to the school's requirement to remain there. Contractors, therefore, must deal with a mix of students in the school district and may not exclude some groups because of the educational and behavioral challenges they may pose.

Evaluation of School Performance in Contracting

The school board could hire a service provider as an intermediary organization to evaluate the contract school. This would enable the school district to assure parents that student needs are being met through services provided and that educational services are of good quality. If parents become dissatisfied because their expectations of the school have not been met, the school contract would be terminated and students would be moved to better schools. This is a great responsibility for the school board. They would have to establish clear expectations by which to evaluate schools. They would have to provide definite evidence that a school is not meeting those expectations. And, they would have to have access to other service providers so that a new contractor could be selected.

The school board, through this process, would be able to raise student performance in all its schools. Competition would be the cornerstone of school improvement. This would only be possible if everyone in the school system recognized and supported the goals and expectations set. Parents, teachers, and students clearly informed of the school's articulated mission could work together to achieve it.

Policy Making in Contracting

The school board would act as a policy making body on a school-by-school basis. Curriculum issues would be dealt with school-by-school, based on each school community's beliefs. Schools, therefore, might differ considerably in services and programs provided. The contract provider, rather than individual school staff members, would be accountable to the school board for goods and services. Effectiveness would be judged on the basis of school designed expectations, which may include compliance, student performance, long-term goal achievement, and family satisfaction. School boards would pay special attention to a contracting school's record of student performance, goal achievement, and family satisfaction. Accomplishment in these area respond to parental concerns and to the business community's expectations concerning the teaching of democratic values and basic employment skills.

The Contractor's Role in Contract Schools

The independent contractor is accountable to the local school board for providing satisfaction and to their own management group for reputation and fiscal responsibility. Schools may close and contracts terminate if contractors do not deliver promised services. The school board must have alternative service providers ready to take control of failing schools and move forward. The primary purpose of the contractor must be to assure contract compliance in the areas of instruction, student outcomes, graduate performance, and contract operations. The school board remains ultimately responsible for the delivery of services and must, therefore, establish the bases upon which satisfaction with the provider will be measured. Expectations should be realistic and provide for ongoing monitoring. School boards must also align their levels of satisfaction with existing evaluative criteria imposed by federal and state guidelines. Contracting in this manner provides the school board with a clear path to student improvement through accountability.

SCHOOL CHOICE AND STUDENT PERFORMANCE IN PUBLIC SCHOOLS

Witte's paper discusses the impact of choice options as a viable means of improving school systems that would lead to increased student achievement. Witte defines public school choice as "any arrangement which deviates in assigning pupils to schools from one based exclusively on geographic location of students to one which provides for some degree of family choice in school assignment." Witte does not define student achievement, although he states that it must be "broadly defined." School choice was not an issue in the traditional school setting when community supported only one school, in which enrollments were organized by age groups. School that did not fit this mode were considered special and created for a specific purpose. Public education in this situation was based on the needs of the communities where students lived.

In the 1960s, the school choice movement was used to integrate large city schools without busing as minorities were attracted to those schools thought to have higher quality than others. This changed the ethnic mix of students in city schools, but had little effect on increasing student performance. In his description of this early school choice movement, Witte raises an interesting issue. As the staffs in choice schools will need to attract clientele on the basis of parent satisfaction with teaching and learning, should school choice be primarily identified with increasing student performance? Is school choice, therefore, a means to improve the current educational system?

Aligning School Choice With Performance

Choice programs delineated in Witte's paper include interdistrict choice programs, open enrollment, charter schools and magnet schools. Several studies were cited that discuss choice options as they relate to performance. In all cases, even though student and parent attitudes toward programs and teachers changed positively, no increase in student performance was indicated. What did occur in the choice option schools was an academically and ethnically mixed student body. Witte found, however, that the majority of students remained in their geographically assigned schools, even with the choice option available.

Although studies were conducted on magnet school effects over a period of years, it was difficult to discern the real impact the organization had on performance, as opposed to other factors, such as parental involvement, that may have influenced results. It is very clear that school choice as a means of improving the current public school system is an important urban issue. Choice options appear to offer an alternative means to make public school programs more relevant to a community's youth and families.

The final choice options Witte discusses are charter schools and open enrollment. The charter school option provides for the creation of a school with a specific mission or curriculum focus, governed by specific state policies. Open enrollment permits students to enroll in any school district in the state, if transportation is provided by the student. No indicators of performance for charter schools or open enrollment were obtained, but ideas for changing the current public school image are evident in both options.

There is much information available on choice programs as they relate to public school organizations. They appear to be expanding as a logical alternative to the existing public education system. No significant findings have been obtained, however, as to school choice improving student performance, which is equated to improving student and parent satisfaction with the system. Perhaps choice programs need to be evaluated by new measures. Perhaps the variables we use to measure student performance need changing to obtain more information specifically on characteristics that lead to satisfaction with the organization.

IMPLICATIONS FOR PRACTICE

Hill's paper and Witte's paper both discuss school reform and accountability as a means to improve student achievement centering on the school house as the unit of change; and they both, therefore, offer several suggestions for characterizing good schooling. They both feel that the school is an organization that has the ability to change itself through enrolling students who elect to attend, and parents, therefore, must see it as a place where they and other community members set expectations and anticipate that they will be met. Clear mission, clear expectations for student performance, parent support for what happens, and autonomy for staff and administration at the school site are success factors for schools that will keep them in business.

Each school house would be competitive in nature and vie for enrollment of students. Instructional practices would align with student needs, as student and community satisfaction is the key to sustaining worth. And, the school staff would have a vested interest in seeing that the school satisfies its customers because staff employment would be contingent on that satisfaction.

The concept of each school house being different is also crucial. The difference may occur through a magnet school, centered on an individual content area; or it may be a charter school that seeks to instruct students through a non-traditional approach. Whatever the theme, the school must be aligned to the community's values and made accountable to that community.

Professional development for staff would be the responsibility of the school and based on staff and student needs. Each school would focus on its own mission and instructional objectives. The premise is that with competition for students, each staff member has a vested interest in making the school successful and, therefore, would be willing to collaborate to ensure that success. Success in this circumstance does not have an overriding definition. It is defined and supported by the stakeholders. They are in the best position to quantify successful student performance from parent, community member, local business, and student perspectives.

The ability of students and parents to choose their school and to have it equitably funded and invested with a quality instructional program could launch a reform of the system that meets the community's definition of satisfaction. The school house would be the center of that community. The larger system would have little purpose other than to monitor the contracting of the service provider to assure that they fulfill their obligations. As with the kaleidoscope, the slightest turn provides a new picture of the current image. With each change in schooling a new pattern forms, bringing another image that could lead to improved learner achievement and higher community satisfaction.

RECOMMENDATIONS FOR FURTHER STUDY

Further study is recommended in eight areas:

1. The definition of satisfaction as it relates to student performance or student/parent expectations needs delineation and quantification. Hill's and Witte's papers do not explain how performance or satisfaction would be measured, analyzed, and determined to meet expectations. There is also the question of whose level of satisfaction must be met by the school to sustain worth. Would it be that of the parents, the students, the contractor, the school board, or all?
2. The professional growth responsibility for sustaining quality rests solely with the school. Discussions in the Hill and Witte papers do not go far enough in suggesting theories on changing the current system or the training needed that will lead to a meeting all student needs. The school will be different, but in what way? Who will model change for the little red school house, and who will orchestrate a plan for reaching its goals?
3. The discussions focusing on attendance choice needs further study. What happens to those students who find they do not fit or whose teachers determine they do not fit in a school once they are there? Will the system keep providing additional choices until a fit occurs? What happens to students if their school of choice closes?
4. As selection is based on choice, the school will be responsible for meeting the needs of all students. How will the funding process, furnished directly to the site, provide for services to special needs children? This question concerns not only curriculum needs, but also many infrastructure needs.
5. As both Hill and Witte focus on the school as the primary education provider, further study should examine the impact of other government, social service, and business entities on the education system. The concept

of choice could extend to these agencies, and they could serve as partners in meeting student needs. The school of choice will be one where families and the community at large is engaged in an ongoing conversation about teaching and learning.

6. Educational leaders currently in the school house may not be capable of changing the system. The concept of student choice based on competition or school contracting based on outside leadership suggest that internal leadership cannot improve the organization at the site level. Further studies are needed about characteristics of successful leadership.
7. As funding would be equitably given to each site based on a per pupil allotment, the system of staffing would need to change. Hiring would be competitive and salary structures would not be based on seniority and education, but instead on keeping the school competitive.
8. More information must be gathered on the impact that competitiveness has on school improvement. If schools must survive on the basis of numbers wanting to attend, then everyone connected to the school must strive for excellence in meeting the expected needs. Even in the business world, however, competition does not always lead to improvement. Improvement is predicated on the knowledge of which characteristics in an organization lead to success of those serving and being served. What is being assumed by the issue of school choice is that by giving parents a choice in the selection of their child's school, public schools will become more competitive and thus improve.

CONCLUSION

The concept of school centered reform suggests recreation of the existing pattern by forming new connections to the learners and their families. This, in turn, creates new expectations for schools. This kaleidoscope view of school reform makes sense since the old patterns must change. The new school-centered image must be crafted, reviewed, and sanctioned by the students, teachers, families, and community.

For many students, parents, educators, and community members, the thought of even gently turning the kaleidoscope is scary, for it means that expectations must change to meet the new patterns formed. Some would prefer to stay with existing conditions because they are predictable and they have served well and they are safe. In many cases they are touted as exemplary. New opportunities, however, must continually be found to improve the current organization. Whether we look to school choice, contract schools, magnet schools, or charter schools, all must provide opportunities for stakeholders to talk about their school and to set expectations that lead to satisfaction. Whatever choice option is selected, the process must begin.

SEEKING CLARITY IN THINKING ABOUT SCHOOL CHOICE AND SCHOOL-CENTERED REFORM

Richard Kamm

SCHOOL CHOICE AND STUDENT PERFORMANCE IN PUBLIC SCHOOL

Witte poses two interesting and important questions in his paper. His first question asks what we know about the effects on student performance of four organizational arrangements for providing public school choice. He concludes that "what we know is more positive than negative," but qualifies his statement later in the paper by saying, "The effects on achievement are somewhat thin."

Considering his discussion of the studies cited in the paper, evidence that choice programs have a positive effect on student achievement indeed seems thin. For three of the four forms of choice discussed (interdistrict choice, open enrollment, and charter schools) no evidence of the effects on student achievement is presented. Evidently, no assessment of these forms of choice has been made. This is understandable in the case of charter schools, but less

Advances in Educational Administration, Volume 5, pages 219-228.

ISBN: 0-7623-0081-7

so in the case of interdistrict choice and open enrollment. Charter schools are a recent development. Very few have been in operation for more than a year or two, but the other two options have been around for some time. The form of choice for which there is some evidence is magnet schools.

Witte discusses nine magnet school studies; however, only three would seem to supply any solid evidence of a positive effect on student achievement. The most sophisticated, according to Witte, is the Gamoran (1996) study. Only the effects on tenth grade students are presented, however, and after within school effects were accounted for, positive effects remained only for reading and social studies. The Witte and Walsh (1990) study of Milwaukee magnet schools also showed significant positive effects for high schools, but Witte cites several limitations in the study, including the lack of an adequate test for selectivity. The Kansas City studies showed positive effects for magnet elementary schools according to Orfield and Eaton (1996), but again Witte points out serious limitations.

An important and surprising aspect of the evaluation of the effect of magnet schools is who attends them. According to Witte, evidence from the Driscoll (1993); Gamoran (1996); and Plank, Schiller, Schneider and Coleman (1993) studies indicates that African American, Hispanic, and lower socioeconomic status (SES) children are more likely to attend magnet schools than "other students." I would like to know more about this, particularly the definition of other students. Does that category include all students in a catchment area who attend public and private schools, or does it refer to the public school population? If the "other" category includes all students, the finding hardly seems surprising, particularly since the magnet school arrangement is more common in urban areas and the studies seem to focus on them. The more affluent and probably more able students in these urban areas may choose private and parochial schools. On the other hand, if African American, Hispanic, and low SES students are more likely to attend urban magnet schools than their public school counterparts, it would seem worth finding out why.

A second question Witte poses, but does not attempt to answer, is: "Why do we know less, and experts seem less concerned, about public school choice than private school choice?" As he points out, these choice programs have associated costs and have given rise to considerable turmoil. The lack of any effort to design ways of determining the effects of public school choice programs, particularly those that are large scale, is indeed puzzling. Although it is apparent that, for the most part, they were designed to answer political rather than educational problems, that does not preclude an effort to determine their effects on student achievement.

In presenting his system for classifying the major public school choice options, Witte lists what he believes was, or is, the motivation for that option. He posits that in the 1980s the motivation for magnet schools, most of which are in large or medium sized districts, was choice and innovation. I suggest that a more powerful factor was school systems wanting to prevent the flight of the more affluent students to parochial schools and private schools and to public schools in suburban districts. Furthermore, I believe this is what is motivating many school systems' current efforts to develop theme schools or magnet schools. School systems do not want to lose the children of community elites. Their children tend to boost test score means, and they are the movers and shakers who can help pass tax referenda.

I suppose it is understandable that when faced with major problems such as achieving racial integration and preventing the flight of community elites, legislatures and local school boards simply plowed ahead with any option that seemed to hold some promise of alleviating the political problem, even though the option was untested. They also didn't bother to take measures to assess the impact of the option on student performance. It is a little more difficult to comprehend why twenty-three states have adopted charter school laws without any evidence that charter schools will produce improved student performance or other positive results. The political pressure for charter schools seems far less intense than the pressures that resulted in interdistrict choice and magnet schools. Given the risk involved in turning over the education of a potentially large number of children to school operators with no solid record of success, it would seem prudent to provide for a thorough longitudinal evaluation. To the best of my knowledge, however, none of the charter school laws provides for the development and funding of an effective longitudinal evaluation.

I hope that the work of Witte and others will serve as a wake-up call to legislatures and school districts, pointing out that programs such as these may result in considerable cost and turmoil and that they should be pilot tested and their effects assessed prior to wholesale adoption. United States schools have been criticized for readily adopting faddish innovations without any real evidence that they work. This was probably a just criticism. It seems that legislatures and school districts do the same when it comes to organizational arrangements. The adoption of organizational changes that impact large numbers of students should be followed by serious efforts to answer the excellent questions Witte poses at the end of his paper: "What are the overall effects on district achievement?" "Who is gaining and who is declining?" "What are the effects on the attitude and morale of teachers both in and not in charter or magnet schools?" I believe another important question should be asked: What is the effect of magnet and charter schools on the allocation of teachers among schools?

The question regarding the effect on the attitude of teachers and administrators is one with which I have had some experience in the magnet school context. In a district with which I worked for four years to improve student achievement, teachers in non-magnet schools consistently expressed feelings that the best students had been creamed off. As one magnet school used test scores as the principal admission criterion and the other used teacher judgment of "student fit," the feelings of those teachers were probably justified.

Their expectations for the students who remained in their schools seemed very low. In the particular school where my efforts were concentrated, I believe that low teacher expectations were a major barrier to improving student achievement. Teachers also felt that the magnet schools were given a disproportionate share of available resources. Although I saw no evidence that their suspicions were justified, this perception impacted their attitude. This is what Witte might call a "not very glorious story;" however, it does support the importance of the questions he posed.

It seems to me that there are additional questions that Witte does not pose. If there is solid evidence that one or more of the choice options result in significant gains in student achievement or other desirable results after controlling for selection bias, parent interest, and so on, what factors contributed to the gains? Do choice schools attract more effective teachers? Are choice schools' curricula more demanding? Are student attitudes toward learning and motivation positively influenced by the ability to choose? The latter seems a strong possibility, at least in the case of high school students. It may explain, at least in part, the Witte and Walsh (1990) finding that the effects in Milwaukee magnet schools were considerably higher for high schools.

The charter school movement, which is just getting underway, seems destined to expand to considerable proportions. It would seem to be a perfect candidate for a major effort to convince legislators to devote some resources to a long range effort to document the effects of charter schools on desired outcomes and to compare them with the outcomes achieved by regular schools. The charter school idea deserves to be tried, if only for the pressure it will bring on school districts to change. The charter school movement does not seem to be causing a great deal of turmoil, and that should make it easier to implement an effective assessment plan. It is early enough in the game that a plan for systematic data collection could be put in place, at least in some locations, before opportunities are lost. I hope that Witte and his colleagues who are studying the effects of choice options can find support for a major effort to find answers to the questions he poses.

SCHOOL-CENTERED REFORM AND ACCOUNTABILITY

I commend Hill for the breadth of his discussion of school choice through contracting. He has touched on most of the important aspects of decentralizing school systems in order to place the locus of control and accountability on the individual school. He does not discuss the likelihood that his contracting plan can be implemented, given the current public school governance structure and the political influence of teacher unions. He foresees his contracting scheme remedying many of the alleged ills of the current system, but he is rather forthright regarding problems that may arise. He foresees contracting resulting in increased innovation, improved student achievement, varied curricula that will meet the needs of diverse populations, increased teacher motivation, more efficient use of resources, greater equalization of resources, and less politicization. He also recognizes the measurement problem and a number of potential problems: teaching to the test, a narrowed curriculum, misleading advertising by schools, board of directors conflict, and attracting teachers to schools in less desirable locations.

I am in favor of a good, fair test of contracting as a means of improving the schooling available to our children because I believe that at least some of the benefits Hill describes could be achieved if all aspects of contracting out public schooling are carefully thought through and the necessary legislation, rules, and regulations are carefully designed. It is also possible that even a pilot test of the Hill plan in one or two states would give the current system a jolt sufficient to cause significant positive change, although charter schools may be a better way of accomplishing that. My response, in which I challenge some of Hill's assertions and raise a number of questions, is offered in the hope that it will contribute to a realistic analysis of the potential results of public school contracting.

I tend to favor a pilot test of contracting, including a rigorous longitudinal evaluation, in a few representative urban areas. This would take several years, but it would seem that a change in the way we provide public schooling as major as Hill proposes should be thoroughly tested and debugged before wholesale implementation, although history does not give me confidence that state legislatures would approach such a change in a rational manner.

Some may contend that contracting was tested in Baltimore, where the school district contracted with Education Alternatives, Inc. (EAI) to manage nine schools, and failed. It was not even close to a test of contracting as Hill envisions it. Under their contract, EAI did not have full control of the curriculum, had to use teachers from the system working under an existing union contract, and did not have full control of all funds. Parents did not have choice. Furthermore, evidence from the University of Maryland group's evaluation seems to indicate that EAI had neither a curriculum nor an effective plan for changing teaching practice.

What about Hill's proposal? Hill states that his so-called "school-centered reform" changes the community to whom schools are accountable, replacing a system based on reporting upward within an administrative hierarchy with a more complex system in which community-wide goals for student achievement, professional standards, and family preferences all matter. But what community? The community served by the individual school the community served by the school district, or some broader community? I suspect he means the one served by the school district. But what about the interests of the broader community, the state's interests? Hill does not explain why community-wide goals and family preferences are not reflected in the current system of local control of public schools. He also does not explain why his new system is more complex. I find it interesting that at the same time Hill is calling for individual schools to be able to develop their curriculum and "approach to instruction," others are calling for a national curriculum and national standards. The latter attempt to make their case by citing the inability of states and local districts to develop clear goals and standards, a high student mobility rate, and the need for a core of common knowledge.

Hirsch (1996) makes a very effective case for a common core curriculum for all schools. He describes the belief that local school districts or individual schools have a coherent curriculum as a myth. Could individual schools, rather small units, do a better job?

How does each community define its standards? Who in the community defines them? It seems to me that evidence from the annual Phi Delta Kappa Gallup Poll and other surveys indicate that one of the current problems with the public school system is that some communities have very low standards and are apparently satisfied with rather poor academic achievement. It also seems that some communities have high standards only for such areas of student performance as athletics or marching band. I don't believe contracting, in and of itself, would raise communities' standards for student academic performance, and I tend to believe communities will not get schools that produce a high level of academic achievement until they demand them.

Hill envisions each school as an independent organization, with its own staff, mission, and approach to education. He places a great deal of emphasis on the benefits of decentralization, which he contends will result in increased innovation, diverse approaches to instruction, less regulation and bureaucracy, and increased teacher motivation. Incidentally, I wish that he had defined what he means by approaches to instruction. Decentralization may have some or all of these benefits, although it is clear that such benefits are not automatic and may be overstated. There is very little evidence that previous decentralization plans have improved student performance.

It seems clear that Hill has no objection to a local school board contracting with a for-profit organization for the operation of a school. It is less clear, however, whether it is acceptable for local boards to contract with organizations that would run multiple schools. For example, could a local board contract with the Edison project, a for-profit corporate organization, for the operation of several or all local schools? If so, how does the fact that Edison schools have a standard curriculum with standard texts, standard organizational formats, and a uniform set of achievement standards fit with his concept of decentralized school centered reform? Might the result be trading one form of bureaucracy for another? Hill suggests that intermediary organizations could provide economies of scale in designing curriculum and staff development, but isn't this likely to lead us back to the kind of centralization and one size fits all schools he decries at other points in his paper?

One wonders if, in the long run, Hill's proposal isn't likely to result in a few well-run and well-capitalized intermediary organizations operating most, if not all, of the schools where his form of contracting is permitted. That is certainly what has happened in many segments of the private sector. It is also what has happened in the case of those services that schools currently contract out. When schools began contracting for student transportation services, food service, and custodial service, there were a large number of purveyors, and it was common for a reasonable number of firms to respond to bid solicitations. Now, many districts often find it difficult to locate a bidder to compete with the current purveyor. A few large firms dominate the market. Maybe they deserve to, but when it comes to the operation of the educational program, is such a result acceptable to Hill, and if not, how can it be prevented?

Hill states that individual schools will be responsible to their own boards. He doesn't elaborate on the composition or role of such boards, but isn't this another governance layer that creates the same opportunities for micro-management that he is trying to eliminate? He contends that his plan will free schools from micro-management by political bodies and protect schools from political turbulence; however, he acknowledges that it will not eliminate political pressures. Contracting may provide school boards a plausible method for managing such political pressures, but it is by no means certain that boards of education will use that method. I do not see how contracting, as he describes it, will prevent interest groups from effectively pressuring legislatures for their pet wants, whether they be curricular mandates, a specific type of pedagogy, uniform requirements that would work against innovation, warrants to be different, or the diversity of approaches to schooling that might alleviate local political pressures. Nor do I believe it will completely insulate individual schools from meddling by district boards. It is the district board, after all, that must approve individual schools' contracts.

Hill contends that under his plan policy would be made on a school-by-school basis, but to what extent is that possible? When a number of schools are the final responsibility of one board of education, aren't there many areas in which policy differences would cause political problems? For example, will it be acceptable to school district voters for individual schools in the same district to have different policies with respect to student discipline, student attendance, or graduation requirements? I contend that given the current school governance system, parents, other community members, and vested interest groups will continue to influence school policy and practice through their district boards much as they do now.

An interesting question is whether Hill would require boards of education to contract out the operation of the schools or simply grant them the authority to do so. If boards are given the authority to contact for the operation of schools, which it appears they already have in some states, will there be any contract schools? Those states that have passed charter school laws limiting the power to grant charters to local boards are finding that very few, if any, charters are granted. When Hill talks about insiders who are reluctant to give full control to individual schools, he must remember that school board members are insiders. I believe it will take considerable public pressure and a legislative mandate to cause boards to contract for the operation of all their schools.

Sarason (1996) makes a compelling argument that we will never have effective school reform in the United States as long as we operate schools under the current governance structure, and Hill's proposal does nothing to change the basic governance structure. This is not the place to examine Sarason's thesis, but it is worth careful study before concluding that contracting will work to reform the system as Hill believes it would.

Two of Hill's major assumptions are that parents will have an exercisable choice of schools and that they will make their choices on the basis of the criteria he deems important. It seems to me that parents' choice of schools will always be limited by geography, access to transportation, and parents' time and financial resources. After all, there will not be a school on every corner. Urban schools, in particular, tend to be large and serve sizable catchment areas. For the urban poor the cost of transportation and the tendency to feel uncomfortable outside their neighborhoods will be limiting factors. Furthermore, it is easy to say that boards can terminate their contacts with failing schools and replace them with good contractors, but will good contractors always be available and interested in operating schools that serve the poor? Will district boards have the political courage to close schools that are producing subpar academic achievement but have satisfied parents?

A key question is: If given the opportunity to choose, on what will parents and students base their choices? Evidence from the Minnesota state-wide choice program indicates that the principal factor in the decisions of parents who exercise school choice is the location of the chosen school. Convenience, particularly when both parents are employed, may be the major criteria. This seems particularly likely in the case of parents with young children. Hill seems to believe their choices will be based on schools' track records with respect to producing high levels of academic achievement.

Evidence from the annual Phi Delta Kappa Gallup Poll, however, indicates that while parents are critical of public schools as a whole, they tend to view their children's schools rather favorably. In the November 1995 poll, 67 percent of parents gave their child's school a grade of A or B, and only 8 percent failed it. In customer satisfaction surveys of parents in six urban-type Illinois schools conducted by the Ball Foundation of Glen Ellyn, Illinois, parents in all six were very satisfied with the schools, in spite of the fact that in four of the schools the scores on statewide reading and math tests were well below the state mean for the five years prior to the survey. I fail to see how contracting would necessarily raise parents' expectations with respect to the quality of the academic programs provided by their children's schools.

I hope that the quality of the academic program would be a major factor in all parents' school choice decisions, but there seems little evidence to support the conclusion that it would be the norm. It may be a reasonable expectation in the case of affluent, well educated parents. Unfortunately, those who most need to choose a good school for their children, the poor and uneducated, may be the least able to evaluate schools and make decisions that are in their children's best interests. They would certainly seem to be the most susceptible to misleading and deceptive advertising. They might also tend to be heavily influenced by aspects of the school program, such as winning athletic teams and flashy marching bands, which are peripheral to the school's central mission.

In one of his tables, Hill indicates that one of the problems with the current system is that school-specific information is seldom available. The implication is that district boards do not have the information they need to take action to fix or to close schools producing low academic achievement, and parents lack the information they need to make informed school choices, even if they had a choice. This may be true in some areas, but I have doubts. Every standardized test report that I have seen shows scores on a school by school basis as well as on a district wide basis. In Illinois, every individual school must annually mail each parent a school report card that provides detailed information on the school's performance on state achievement tests. The *Chicago Tribune* annually publishes the state test scores for individual schools and has made the scores for every Illinois school available on its Web page. This reporting requirement has been in effect for several years, but it has had little impact on school districts or individual schools. It seems the problem is a lack of high expectations by demanding school customers, rather than a lack of information.

One of the most attractive aspects of Hill's plan is that it puts the focus on the individual school as a unit, a unit with the authority to make programmatic and resource allocation decisions and to be held accountable for the results. The school seems a reasonably-sized unit for developing a team of teachers, administrators, and support staff who collaborate to meet the educational needs of their students. It is small enough that it may be possible to make many decisions on the basis of consensus and to develop a sense of ownership and community among students and staff.

The extent to which school level accountability results in increased staff motivation to excel probably depends, in large part, on good, fair performance measures. Hill's focus on a value-added approach to assessing a school's effect on achievement is theoretically sound; however, the value-added approach is neither easy nor inexpensive.

Hill also suggests the formation of an independent "inspectorate" that would rate the quality of schools. As the Illinois State Board of Education has learned, developing appropriate criteria and an effective inspection system is a difficult task, and assembling, training, and supporting a qualified cadre of inspectors is difficult and costly.

Hill's suggestion that the measurement of ultimate results should be an accountability mechanism sounds appealing; however, it also presents some challenges. Following students in our highly mobile society is difficult, as many of us who have attempted follow-up studies of high school graduates have learned. Furthermore, so many other variables affect graduates' occupations and employment status that selecting out school effects seems close to impossible. It does, however, seem worth some effort to attempt to find measures of students' ability to engage the next level of schooling and graduates' performance on real world tasks.

In sum, Hill has provided the broad outlines of an approach to decentralizing the operation of individual schools on a large scale. He has by no means, however, analyzed the possible problems and pitfalls. He has not touched on the legislative changes, particularly with respect to collective bargaining, that will be necessary to insure that another of the best laid plans of mice and men does not result in yet another confirmation of the maxim that the more things change, the more they stay the same. Hill has not dealt with the problem of helping parents become the discerning, demanding consumers of schooling necessary to make contracting, or any other choice plan, an effective means of improving student academic performance.

REFERENCES

Driscoll, M. E. (1993). Choice, achievement, and school community. In E. Rassell & R. Rothstein (Eds.), *School choice: Examining the evidence* (pp. 147-172). Washington, DC: Economic Policy Institute.

Gamaron, A. (1996). Student achievement in public magnet, public comprehensive, and private city high schools. *Educational Evaluation and Policy Analysis, 18* (1), 1-18.

Hirsch, E. D., Jr. (1996). *The schools we need.* New York: Doubleday.

Orfield, G., & Eaton, S. E. (1996). *Dismantling desegregation: The quiet reversal of Brown v. Board of Education.* New York: The New Press.

Plank, S., Schiller, K. S., Schneider, B., & Coleman, J. S. (1993). Effects of choice in education. In E. Rassell & R. Rothstein (Eds.), *School Choice: Examining the Evidence* (pp. 111-134). Washington, DC: Economic Policy Institute.

Sarason, S. B. (1996). *Parental involvement and the political principle: Why the existing governing structure of schools should be abolished.* San Francisco: Jossey-Bass.

Witte, J. F., & Walsh, D. J. (1990). A systematic test of the effective schools model. *Educational Evaluation and Policy Analysis, 12* (2), 188-212.

WHO BEARS THE BURDEN OF SCHOOL ACCOUNTABILITY?

Paul W. Thurston

Paul Hill is a seminal thinker in encouraging public schools to privately contract for services necessary to deliver student instruction. In his 1994 Rand monograph, *Reinventing Public Education*, he proposed contracting as a new way of governing public education. Simply stated, a district school board would contract with independent organizations to operate particular schools. Each school's contract would specify the public funds it would receive, the type of instructional programs it would offer, and the expected student outcomes. Under this system, Hill envisions the school board contracting with a variety of educational vendors for a range of different types of schools. The contractors, rather than the district, would employ the teachers. The major responsibility of the school board would be to bid, to negotiate, and to enforce contracts.

In his paper in this book, "Accountability under Charters and Other School Centered Reforms," Hill continues his call for contracting as a way to bring reform and vitality to public education. As in his earlier work, Hill sees contracting as going beyond the service activities in which schools engage, such as janitorial or bus services, and he calls for contracting the core educational

Advances in Educational Administration, Volume 5, pages 229-235.

ISBN: 0-7623-0081-7

activities—teaching and delivery of the instructional program. Central to this focus on contracting is Hill's commitment to the school as the central organization through which education is delivered. Schools, not school districts, deliver instruction. It is the school, therefore, that is the unit of analysis in contracting.

Hill's attention is focused on the school site. Public school contracting, where an independent organization contracts with the school board to operate a particular school, is but one mechanism for delivering a variety of distinctive schools to which families could choose to send their children. Other mechanisms include charter schools and decentralized schools. Charter schools are supported by state funds and allow the formation of private entities to contract with the state according to the standards provided in the charter school legislation. Decentralized schools are those schools operated by the district that provide particular choices to students and parents. Hill describes them as schools that are given broad latitude to operate outside of the district bureaucracy with discretion over spending and emphasis on student performance rather than regulatory compliance.

Whether the magnet schools and choice schools that Witte (1996) analyzes would satisfy Hill's conception of decentralized schools is not clear. The way Hill lumps these three different types of schools together implies that he doesn't think the mechanism for creating the schools will significantly influence the types of schools that are actually created. The accuracy of this implication deserves considerably more attention.

As in Hill's Rand monograph (1994), the role of the school district is specialized. The school district, presumably the school board, is responsible for providing and maintaining school facilities and for hiring and monitoring the performance of the contractors who operate the schools. The school board will contract for the operation of the school. The contractor will have responsibility for operating the school and meeting the terms of the contract. The arrangement will remove, in Hill's view, any potential of the school board to micromanage the school.

How appealing is the new system of public education that Hill has described as a means of realizing higher levels of student academic achievement, compared to the public schools as presently configured? Hill's clear focus on the individual school as the organization that delivers education is attractive, but the details surrounding realization of a higher performing school are left vague. The variety of delivery mechanisms for creating the individual school masks a central issue of what the individual schools will look like. Different structural mechanisms will provide different schools, and it is not so much a matter of accountability as it is governance that will determine how these schools will operate.

Hill proposes that school districts play a part that is very different and much narrower than their present governance role. He suggests that they should become contract negotiators and performance monitors. He elaborates on this new role of the school district to demonstrate how a contract relationship might operate. Because state legislation would be necessary to authorize this different role and because such a major change in the delivery of public education would undergo intense public debate and legislative scrutiny, why did he not consider alternative superstructures for managing school building contracts? Or, if he did consider alternatives, why did he dismiss them?

School districts, as currently configured in Illinois, exist primarily as historical anachronisms. Their most significant function is to provide the boundaries that define the property used for measuring local district wealth; that definition, combined with state moneys allocated through the state aid formula, largely determine the level of expenditures to support each student. Because of the inequity that exists in the distribution of these resources, I can't help but wonder, at least for Illinois, whether the passage of legislation necessary for contracting wouldn't provide an opportunity to restructure school funding.

For Hill's purposes, there is nothing compelling about the need to maintain school district boundaries in their current form. An alternative proposal would be to draw these school districts with much larger boundaries and with the objectives of obtaining general financial equity across the districts and of providing sufficient scale to accomplish a high level of efficiency. Efficiency at this level would maximize the expenditures that go to individual schools.

Hill's formulation of the school district is so radically different from the current legal entity, that I wonder why, given the emphasis on the individual school site, school districts are even necessary. Perhaps regional offices of education could assume oversight responsibilities for schools, or possibly the state board of education could be reconstituted to deal with local schools. Both of these suggestions have drawbacks, of course, because of the revolutionary changes they would require in the operation of regional offices and the state board. The core of Hill's proposal, the centrality of the school as the delivery unit for education and the potential for contracting for the operation of the school, does not depend on the existence of the school district. It is intriguing, therefore, to explore the implications of governance, funding, and oversight for individual schools for which there is no district structure. If Hill desires to maintain school districts because of citizen involvement in their governance on school boards, how realistic it is to think of their governance responsibilities as primarily contractual? It is hard to think that at least some school board members would not want to become involved in the micromanagement of some schools. Hill's reformulation of the role played by the school board is so radical as to be somewhere between unrealistic and naive.

It is not clear from Hill's description of the relationship between the school and the contractor why anyone would want to become a contractor. There is serious doubt about the extent of autonomy that a contractor would have to operate a school. How much of a warranty for specific levels of student performance can appropriately be required of a contractor? At what point do parents have the option to remove children from the school, and what effect will this have on the payment level to the contractor? Beyond this, there are certain legal benefits that accrue to public school teachers that might not be available to teachers privately contracted. For example, public school teachers enjoy limitations on their personal liability that would involve some financial cost for contractors to replicate. It is possible that the burden of accountability will fall to the contractor. On the other hand, the example of accountability in America's space and defense industries, which shows the compatibility of accountability with complexity and technical uncertainty, may make potential contractors more sanguine when thinking about the potential for cost overruns as a way of doing business.

THE ADVANTAGES OF SCHOOL CONTRACTING

Can some of the potential advantages of contracting at the school site be implemented in public schools with only modest changes in the authority given to them? Private contracting could, theoretically, be done in public schools under the existing legal framework. Technically, in those states where public sector collective bargaining statutes exist, it would be possible for a school board to negotiate for one or more schools operated by an independent contractor. This could be done as part of a new contract or through mid-term bargaining. The practicality of engaging in such bargaining is obviously remote, and it is doubtful that any but the most desperate superintendents or school boards would engage in this strategy to realize school improvement.

Even though contracting for instructional services at a school site is impractical and may be impossible in the legal framework in which schools currently exist, it is important to note that the prospect of private contracting does have a certain attractiveness. Hill does not pursue the potential for private contracting in current school districts because of his belief that school districts are not interested in this form of experimentation, presumably because of the effects of bureaucracy and the lack of motivation to change.

Private contracting is attractive as a way of experimenting with different school labor configurations to provide instruction. With approximately 80 percent of the cost of public education tied up in labor costs, this is the area where most flexibility exists to utilize resources in different ways. In the October 15, 1996, edition of the *New York Times* (Steinberg), New York City School Superintendent Crews reviewed his first year in office, reflecting that his biggest frustration was having little means by which to hold teachers and principals accountable for student performance. Crews said,

> This is the only system that I have ever been in that almost requires that you simply lean on the sheer force of your own will and personality, as well as the sort of officiousness

> of the position itself, to create change. To some degree, those aren't bad tools to have, but they are not sufficient (p. A13).

New York bureaucracy and contract protection may be in a league of their own, but other districts also provide barriers to making substantive changes. I have observed teachers who were willing to take risks with school governance issues and to engage in extensive staff development. When given the opportunity to think about how to reconfigure the staffing pattern of their school, however, they were unable to consider the financial resources invested in staffing positions in any terms other than that their friends and colleagues were holding these positions. Meaningful discussions of reconfiguring these positions were not possible. I find, therefore, the potential of private contracting worthy of further exploration.

Private contracting may provide interesting, viable alternatives for the delivery of instruction. Having said this, I must admit concerns about the potential abuse of such experimentation. Abuse can come in many forms, such as the reconfiguring of positions to hire less qualified persons or treating schools as potential employers rather than as educational institutions. This potential abuse makes me cautious about the scope of experimentation that can be undertaken in this area. I remain open to the concept of private contracting, but I want it to be limited and carefully evaluated so that we can learn from it.

If it were possible to provide opportunity for private contracting within the current governance structure of a school district, this might be a powerful way to build competition and to provide alternative models for improving student achievement within the system. It is arguable that school district organizations may not be so strictly bound by regulation and bureaucracy that they could not be opened to experimentation, if given the authority to do so. Contracting for instructional services might be somewhat more hybrid in providing accountability measures at the district level, while offering considerable autonomy to the contractor at the building level. This appears to be more like Hill's description of accountability.

Proposals are appropriate for broadening the range of options that can be provided within the public school setting. Consistent with what Witte (1996) describes as the increased move in many districts to encourage magnet schools, choice between schools, and increasing interest in charter schools across the country, legislation ought to allow modest experimentation with contracting. This may include expanded private contracting and modifications to present contracting done in some districts as variations on standard working conditions. The choices may be limited.

Hill's attention to school-centered reform is instructive because of its treatment of the school as the unit of analysis. The school is where the instruction is delivered and the point of connection to parents and children. The contracting system Hill proposes is characterized as being preferable because it eliminates the bureaucratic practices of the district that may detract from attention to individual schools. This concern about the educational bureaucracy may be merited in some districts, but many districts already give great attention to the quality of education provided at the building level.

There is serious question about the extent to which there is parental dissatisfaction with the quality of education being provided in neighborhood schools. Ward's paper in this book takes up this matter systematically. Early anecdotal information from an open enrollment program in Minnesota that allowed parents to move their children from one school to another school across district boundaries suggests that it was done for largely noneducational reasons, such as to get closer to a job, family member, or baby sitter.

An argument for school-centered reform and the contracting of services is that these options will free schools from current statutory and regulatory requirements that limit discretion and creativity. When examined closely, however, concerns about overregulation of schools appear to be overstated. Illinois provides a case in point. The Illinois legislature recently passed legislation that provided a mechanism to obtain waivers from a wide range of state educational legislation and regulations. Although there was considerable interest among school boards in seeking waivers, early analysis of waiver requests suggests that they contained no profound changes in the delivery of educational services (Hosea, Colwell & Thurston, 1996). Waivers were sought from physical education requirements and personnel matters, and some districts even sought to reinstitute corporal punishment. Tenure laws and collective bargaining are beyond the reach of the state's waiver statute.

The contracting system Hill proposes seems to assume a non-collective bargaining employment relationship. This may be an attractive part of the proposal to some observers, but serious questions remain about the quality of personnel that will be employed in these schools. The experience in Arizona, with their relatively loose charter school legislation, revealed that nepotism and the questionable qualifications of teachers employed in certain charter schools became a public issue, and the state board of education subsequently tightened the approval process for obtaining a charter.

At the end of his paper, Hill reminds us of the importance of information and making judgments about the quality of education delivered at the school level:

> Any community attempting such a reform [that tries to focus initiatives and accountability at the school level] without considering how to replace a system of compliance-based accountability with one based on school specific performance will soon discover that it faces questions and choices on which it has no relevant data.

Hill's analysis of accountability provides some guidance about what kind of data and information are important in making judgments about educational quality.

The call for school-centered reform is persuasive. Such reforms within existing school district structures seems possible in many districts. The contracting of school service providers holds some promise, but I am suspicious of its implementation until details of the different set of legal relationships are more clearly delineated.

REFERENCES

Hill, P. T. (1994). *Reinventing Public Education.* Seattle: Rand Institute for Education and Training. (DRU-690-IET/LE/GGF).

Hosea, S., Colwell, B., & Thurston, P. (1996). Increasing school district autonomy through waivers of legislative mandates: The Illinois experience. *West's Education Law Reporter, 107* (2), 443-58.

Steinberg, J. (1996, October 15). Educator settles in for long haul after a year on the job. *New York Times*, p. A13.

Witte, J. F. (1996). *School choice and student performance in public schools.* Paper prepared for the First Edwin O'Leary Symposium on Financial Management at the University of Illinois, Urbana-Champaign, October 16-17.

www.ingramcontent.com/pod-product-compliance
Ingram Content Group UK Ltd.
Pitfield, Milton Keynes, MK11 3LW, UK
UKHW021032041026
15470UKWH00014B/1543

9 780762 300815